Andrew H. Baker

Complete arithmetic, or, Third book of a series of mathematics

Andrew H. Baker

Complete arithmetic, or, Third book of a series of mathematics

ISBN/EAN: 9783741185281

Manufactured in Europe, USA, Canada, Australia, Japa

Cover: Foto ©Andreas Hilbeck / pixelio.de

Manufactured and distributed by brebook publishing software (www.brebook.com)

Andrew H. Baker

Complete arithmetic, or, Third book of a series of mathematics

COMPLETE ARITHMETIC;

OR,

THIRD BOOK

OF A

SERIES OF MATHEMATICS.

BY

ANDREW H. BAKER, A.M., PH.D.,

REVISED AND ENLARGED EDITION.

NEW YORK:
P. O'SHEA, PUBLISHER,
87 BARCLAY STREET.
1880.

CONTENTS.

	PAGE		PAGE
Definitions	7	Inverse Ratio	13
Mathematical Terms	8	Compound Ratio	13
Axioms	9	Simple Equations	14
Signs	9	Equation of Payments	14
Notation and Numeration	10	Averaging Accounts	14
Addition and Subtraction	13	Conjoined Equations and Ratios	145
Multiplication	20	Proportion	149
Division	26	Percentage	152
Problems	35	Commission or Brokerage	152
Factoring	36	Stocks, Bonds, Etc.	158
Common Multiple	39	Insurance	162
Greatest Common Divisor	41	Interest	164
Cancellation	42	Review	171
Fractions	45	Banking	178
Addition and Subtraction of Fractions	48	True Discount	180
Addition of Fractions	50	Exchange	181
Subtraction of Fractions	51	Domestic Exchange	182
Multiplication of Fractions	52	Foreign Exchange	184
Division of Fractions	55	Alligation	195
Complex Fractions	57	Involution and Evolution	200
Decimal Fractions	62	Evolution	203
Addition and Subtraction of Decimals	63	Series of Common Differences	213
Multiplication of Decimals	63	Series of Equal Ratios	217
Division of Decimals	65	Compound Interest	219
Circulating Decimals	68	Annuities	224
Review	70	Mensuration	229
Denominate Numbers	94	Measurement of Surfaces	230
Denominate Numbers and Fractions	118	Measurement of Circles	232
Time	119	Cubic Measure	234
Time and Longitude	120	Answers	238
Practical Examples	124	Metric System	247
Ratio	129	Test Examples	249

Copyright, 1878, 1880, by P. O'SHEA.

Electrotyped by SMITH & McDOUGAL, 82 Beekman St., N. Y.

CAJORI

PREFACE.

IN this book, all the principles of Arithmetic are fully developed, and sufficient examples are given to fix them on the mind.

When a student is very apt and thoroughly understands the PRIMARY LESSONS, he may omit the ELEMENTARY, and immediately take up this book, which is complete in itself.

I have discarded puzzles of every kind, which only perplex the student without advancing him a step in science.

A few simple principles of algebra are introduced, in order to elucidate more clearly, the different functions of interest, the series of equal ratios, and the square and cube root.

Problems in mensuration are also given, the principles of which are derived from Geometry.

Arithmetic is a pure mathematical science, and if its principles are systematically developed, the student will progress with easy and rapid steps, and when he has finished this book, he will discover that he has already so far ascended the hill of science that a retrospect will present to him many beauties which are greatly enhanced when seen in their harmonious relation to each other.

PREFACE.

The importance of the Common Fraction, which is now rendered as simple as an Integral Number, cannot be too highly estimated. It affords the greatest variety of Problems for the general training of the mind. It expresses the Ratio of two numbers, and Ratio is the element of Percentage, and of almost every Arithmetical operation. It greatly abridges the work, especially when Cancellation is applied; it has in a great measure superceded the Decimal and Compound Denominate Number; Avoirdupois Weight is seldom carried below the lb.; and the cwt., qr., and lb. are more simply rendered in lbs. only, and the lbs. are readily reduced to the fraction of a ton. In the measurement of dry-goods, only the yard and the fraction of a yard are used. Troy Weight is seldom reckoned below the oz. and the fraction of the oz. The Apothecary requires the small weights in mixing medicines. Time and Circular Measure cannot be changed; nor is there any attempt made at a change in the great variety of coins and currencies, even by the Metric System.

The Author cannot too highly recommend to the Teacher the use of the Blackboard described on the following page. Great facility in comprehending the combinations and divisions of numbers will be acquired by this method.

BLACKBOARD EXERCISE.

This page represents a blackboard with the numbers as high as 72 painted on its margins.

There is also a box containing slips which will cover two, three, four, etc., as high as 12, and numbered accordingly; one of these the student will take in his hand and apply it to the painted numbers to perform addition or subtraction; thus, begin at 1 and take a slip marked 2, then 1 and 2 are 3, 3 and 2 are 5, 5 and 2 are 7, 7 and 2 are 9, etc., counting at least the left-hand column; then, to perform subtraction, begin at the bottom of the 1st column; thus, 36 minus 2 equals 34, 34−2=32, 32−2=30, 30−2=28, etc., until the top is reached; then taking a slip marked 3, begin with 1 or 2, or first with 1 and then with 2, and return to the top of the column as before, by subtraction; let this exercise be performed with all the slips, and as the larger numbers are taken, continue the additions to the bottom of the 2d column, and return as before.

For multiplication and division first make a chalk mark after every two figures up to 24, and multiply; thus, once 2 are 2, twice 2 are 4, 3 times 2 are 6, 4 times 2 are 8, etc.; then the number of divisions is 12 and each division has 2 numbers; ∴ 12 is contained twice in 24, or 2 is contained 12 times, 2 is contained once in 2, in 4 twice, in 6 three times, in 8 four times, in 10 five times, in 12 six times, etc. When the student is familiar with multiplication and division by 2, let the numbers be separated into 3's, then 4's, etc., and let each be continued for 12 divisions; when all the divisions have been performed according to the steps, beginning with 2 and ending with 12, a multiplication and division table will be made.

REM.—In multiplication the product of any two factors is the same by making either the multiplicand and the other the multiplier; so also in division, the divisor and the quotient may be substituted, as the dividend is the product of the divisor and quotient.

REM.—The numbers, continued up to 144, should be painted on the sides of the board.

COMPLETE ARITHMETIC.

DEFINITIONS.

1. *Arithmetic* is the science of numbe___.

2. A *Unit* is a single thing; as, a book, one dollar, or simply one.

3. A *Number* is a unit or a collection of units; as, one, ten, five books, twenty-five dollars.

4. The numbers used in Arithmetic are all formed by combinations of the ten Arabic characters, called *Figures;* viz., 0, called zero or naught; 1, called one; 2, two; 3, three; 4, four; 5, five; 6, six; 7, seven; 8, eight; 9, nine.

5. Expressing a number either in writing or figures is called *Notation*, and reading the expression is called *Numeration.*

6. When numbers are used without reference to any object, they are called *Abstract Numbers;* as, five, twenty, etc.; but when they are applied to things, they are called *Concrete;* as, one book, ten men, four dollars, etc.

7. When concrete numbers express values of money, weights, measures, time, etc., they are called *Denomi-*

nate Numbers; as, dollars, pounds, shillings, pounds of weight, ounces, hours, minutes, etc.

8. When different denominations of either kind form but one number, it is called a *Compound Number;* as, £4 3s. 6d., 2 lb. 1 oz. 3 pwt. and 2 gr.

9. Numbers of the same order and the same denomination are termed *Like Numbers;* other numbers are termed *Unlike Numbers.*

REM.—Numbers expressing different species of the same genus are unlike, as horses and cows; while the same numbers expressed in the term of the genus are alike, as animals.

MATHEMATICAL TERMS USED IN ARITHMETIC.

1. An affirmative sentence, or anything proposed for consideration, is a *Proposition.*

2. A self-evident proposition is called an *Axiom.*

3. A proposition made evident by a demonstration is called a *Theorem.*

4. When a proposition is used for developing a principle of Arithmetic, it is called a *Problem.*

5. Propositions given merely for solution, in order to impress the principles on the mind, are called *Examples.*

6. An obvious consequence of one or more propositions is called a *Corollary.*

7. An established custom, or an assumption without proof, is called a *Postulate.*

REM. 1 and 1 are 2, 2 and 1 are 3, 3 and 1 are 4, 5 and 2 are 7, 6 and 3 are 9, etc., is the postulate which forms the basis of Arithmetic.

AXIOMS.

1. If equal numbers are added to equal numbers, the sums will be equal.

2. If equal numbers are subtracted from equal numbers, the remainders will be equal.

3. If equals be multiplied by equals, the products will be equal.

4. If equals be divided by equals, the quotients will be equal.

5. If two numbers are each equal to the same number, they are equal to each other.

6. If the same number be added to and subtracted from another number, the latter number will not be changed.

7. If a number be both multiplied and divided by the same number, the former number will not be changed.

8. If two numbers be equally increased or diminished, the difference of the resulting numbers will be the same as the difference of the originals.

9. If two numbers are like parts of equal numbers, they are equal to each other.

10. The whole is greater than any of its parts.

11. The whole is equal to the sum of all its parts.

SIGNS.

1. The sign +, called *plus*, is the sign of addition, and indicates that the number on the right hand is to be added to the one on the left.

2. The sign —, called *minus*, is the sign of subtraction, and indicates that the number on the right is to be subtracted from that on the left.

3. The sign ×, called *into*, is the sign of multiplication, and indicates that the numbers between which it is placed are factors of the same product.

4. The sign ÷, *divided by*, the left-hand number to be divided by the right hand.

5. The sign =, *equal to*, indicates that the numbers between which it is placed are equal.

6. 5^2, 5^3, the 2 and 3 placed to the right, a little above a number, indicates the *power* to which it is to be raised.

7. $\sqrt{}$, $\sqrt[3]{}$, indicate the extraction of the *square* and the *cube root*.

Notation and Numeration.

1st. A figure standing alone, as 1, 2, 3, holds the units place, or is of the 1st order, and is read, *one, two, three*.

2d. A number having two figures, as 14, 26, the right-hand figure holds the units place, and the left-hand figure that of tens, and they are read, *fourteen, twenty-six*.

COR.—The right-hand figure of a number is called *units*, or the 1st order; the next figure to the left is called *tens*, or the 2d order; the third figure, *hundreds*, or the 3d order; the fourth figure, *thousands*, or the 4th order; and if a number be expressed with the nine figures in order, making 1 the right-hand figure, the figures will express their respective orders; thus,

NOTATION AND NUMERATION.

```
           millions,  thousands,   units.
         ⎧   ⎫      ⎧   ⎫       ⎧   ⎫
         hundreds of    hundreds of    hundreds of
             tens of        tens of        tens of
                units of       units of       units of
          9   8   7  ,  6   5   4  ,  3   2   1
```

If pointed in periods of three figures each, they may be read as follows: *Nine hundred and eighty-seven millions six hundred and fifty-four thousand three hundred and twenty-one.*

REM.—The figures designate the orders.

Read the following numbers:

1 . . . . one.
21 . . . . twenty-one.
321 . . . . three hundred and twenty-one.
4,321 . . . . four thousand three hundred and twenty-one.
54,321 . . . . . . fifty-four thousand three hundred and twenty-one.
654,321 . . . six hundred and fifty-four thousand three hundred and twenty-one.
7,654,321 seven millions six hundred and fifty-four thousand three hundred and twenty-one.
87,654,321 { eighty-seven millions six hundred and fifty-four thousand three hundred and twenty-one.
987,654,321 { nine hundred and eighty-seven millions six hundred and fifty-four thousand three hundred and twenty-one.

REM.—The column of 1's is of the 1st order, the column of 2's is of the 2d order, the 3's the 3d order, the 4's the 4th order, etc.

COR.—The relation of any two consecutive orders is the same, for when in addition the sum of any column reaches 10, the left-hand figure belongs to the next column or order; hence, a table may be formed, thus,

```
10 units            = 1 ten.
10 tens             = 1 hundred.
10 hundred          = 1 thousand.
10 thousand         = 1 ten-thousand.
10 ten-thousand     = 1 hundred-thousand.
10 hundred-thousand = 1 million.
       etc.               etc.
```

NOTATION AND NUMERATION.

This method of numeration may be extended; thus,

Septillions.	Sextillions.	Quintillions.	Quadrillions.	Trillions.	Billions.	Millions.	Thousands.	Units.
Hundreds of / Tens of / Units of	Hundreds of / Tens of / Units of	Hundreds of / Tens of / Units of	Hundreds of / Tens of / Units of	Hundreds of / Tens of / Units of	Hundreds of / Tens of / Units of	Hundreds of / Tens of / Units of	Hundreds of / Tens of / Units of	Hundreds of / Tens of / Units of
1 2 3,	4 5 6,	7 8 9,	1 2 3,	4 5 6,	7 8 9,	1 2 3,	4 5 6,	7 8 9

REM.—This is called the French Method of Numeration, and is generally followed; the English Method has six figures in each period, as follows:

of Trillions.   *of Billions.*   *of Millions.*   *of Units.*

Etc.   Etc.   Hundreds of thousands / Tens of thousands / Thousands / Hundreds / Tens / Units   Hundreds of thousands / Tens of thousands / Thousands / Hundreds / Tens / Units

1 2 3 4 5 6, 7 8 9 1 2 3, 4 5 6 7 8 9, 1 2 3 4 5 6

Read the following notations:

1.	123.	8.	245,678,954.
2.	1,234.	9.	365,421,783.
3.	12,345.	10.	204,603,207.
4.	123,456.	11.	100,200,300.
5.	1,234,567.	12.	20,030,040.
6.	12,345,678.	13.	3,004,005.
7.	123,456,789.	14.	12,302,105,401.

ADDITION AND SUBTRACTION.

ADDITION AND SUBTRACTION TABLE.

0	1	2	3	4	5	6	7	8	9
1	2	3	4	5	6	7	8	9	10
2	3	4	5	6	7	8	9	10	11
3	4	5	6	7	8	9	10	11	12
4	5	6	7	8	9	10	11	12	13
5	6	7	8	9	10	11	12	13	14
6	7	8	9	10	11	12	13	14	15
7	8	9	10	11	12	13	14	15	16
8	9	10	11	12	13	14	15	16	17
9	10	11	12	13	14	15	16	17	18
10	11	12	13	14	15	16	17	18	19
11	12	13	14	15	16	17	18	19	20
12	13	14	15	16	17	18	19	20	21
13	14	15	16	17	18	19	20	21	22
14	15	16	17	18	19	20	21	22	23
15	16	17	18	19	20	21	22	23	24
16	17	18	19	20	21	22	23	24	25
17	18	19	20	21	22	23	24	25	26
18	19	20	21	22	23	24	25	26	27
19	20	21	22	23	24	25	26	27	28
20	21	22	23	24	25	26	27	28	29
21	22	23	24	25	26	27	28	29	30

The square made by the ten Arabic characters forms an Addition and Subtraction Table.

Beginning with the first line, thus, Zero and zero are zero; zero and 1 are 1; zero and 2 are 2; zero and 3 are 3; zero and 4 are 4; zero and 5 are 5; zero and 6 are 6, etc.

The second line, one and zero are one; 1 and 1 are 2: 1 and 2 are 3; 1 and 3 are 4; 1 and 4 are 5, etc.

2 and zero are 2; 2 and 1 are 3; 2 and 2 are 4; 2 and 3 are 5; 3 and 4 are 7, etc.

3 and 0 are 3; 3 and 1 are 4; 3 and 2 are 5; 3 and 3 are 6; 3 and 4 are 7, etc.

Continue this, taking the first figure of the 1st column and adding it to each successive figure in the first line; the adding of zero is only nominal, as it makes no increase.

It also becomes a subtraction table, the figures of the first column being the subtrahend, and those of the first line the remainders.

Take zero from 1 and 1 remains; 0 from 2, 2 remain, etc. It may be thus expressed: $1-0=1$; $2-0=2$; $3-0=3$; $4-0=4$; $5-0=5$; which is read, 1 minus zero equals 1, etc.

Second line: take 1 from 2, 1 remains; 1 from 3, 2 remain; or, $2-1=1$; $3-1=2$; $4-1=3$; $5-1=4$, etc.

Third line; $3-2=1$; $4-2=2$; $5-2=3$; $6-2=4$; $7-2=5$, etc.

In addition, we add two numbers at a time, never more, and in the first square we have the addition of every two units that can come together; so also in subtraction.

ADDITION AND SUBTRACTION. 15

In the second square, the units correspond with the first square, and have an additional ten.

In the third square, the units again are repeated, and another additional ten.

As a column of tens, hundreds, and every higher or lower order is added and subtracted in the same way, the above table develops every principle of addition and subtraction.

Add the column of units.

```
1    3
2    5
3    6
4    4
5    2
6    7
7    9
—    —
28   36
```

One and 2 are 3 ; 3 and 3 are 6 ; 6 and 4 are 10 ; 10 and 5 are 15 ; 15 and 6 are 21 ; 21 and 7 are 28 ; or as is customary to begin at the bottom of the column, 7 and 6 are 13 ; 13 and 5 are 18; 18 and 4 are 22; .22 and 3 are 25; 25 and 2 are 27; 27 and 1 are 28.

$9 + 7 = 16$; $16 + 2 = 18$; $18 + 4 = 22$; $22 + 6 = 28$; $28 + 5 = 33$; $33 + 3 = 36$.

REM.—Although many numbers may be added together, in performing the operation only two at a time are added.

Add the following numbers jointly and separately; thus,

```
35  =  30  and  5
24  =  20   "   4
43  =  40   "   3
52  =  50   "   2        21
67  =  60   "   7       200
———    ———      ——      ———
221    200   "  21   =  221
```

The sum of the column of units is 21 ; that is, 1 unit and 2 tens; the sum of the column of tens is 20; that is, 20 tens or 2 hundred; and the two sums united make

221; precisely the same as if the column of units is first added, and the units of the sum placed under the column of units, and the tens added with the column of tens; and then the tens of the sum of the tens column placed under the column of tens, and the hundreds in place of hundreds.

Cor.—As the relation of each successive order is the same, hence for every ten of any order, the 1, or left-hand figure, belongs to the next order; and the process is the same in the addition of every column; that is, one is carried to the next column for every ten in the addition of each column.

ADDITION.

```
 3241    4365   643215
 4356    5321   532684
 6745    7546   478921
 5364    8432   586432
19706
```

```
754321   864321   345678
678643   678963   987654
594721   987654   321987
367543   456789   654321
```

Add

```
 7654
 3897
11551
```

Subtract
   Minuend,   11551
   Subtrahend, 7654
                  3897

SUBTRACTION.

```
876432   987654
543210   321234
```

```
876543   789654
654321   321043
```

```
869754
654321
```

```
3465321
6354789
9820110
```

   Minuend,   11551
   Subtrahend, 3897
                  7654

Minuend,	9820110	*Minuend,*	9820110
Subtrahend,	3465321	*Subtrahend,*	6354789
	6354789		3465321

Cor. 1.—The minuend is always equal to the sum of the subtrahend and remainder, and is therefore greater than either.

Cor. 2.—Arithmetic is based upon the postulate contained in 1, 2, 3, 4, 5, 6, 7, 8, 9, 10, 11, *which is addition,* etc.; and the application of Axiom 6 (page 9) to this postulate proves the principle of subtraction; thus, $19 + 6 = 25$, then $25 - 6$ must be equal to 19.

Rem. 1.—When the figure of the subtrahend is larger than the one above it of the same order of the minuend, 1 of the next order of the minuend must be united to the figure of the minuend and then the subtraction be performed; then in order to make up for this addition to the minuend, 1 must be added to the next order of the subtrahend, and then perform the subtraction; this process is called carrying and requires all the attention of the student.

Rem. 2.—I prefer few examples, but these may be often repeated, and if thought necessary the teacher can give others in which the columns are longer.

Rem. 3.—Each order may be regarded as units, and the sum may reach one, two, or more hundred of its order.

QUESTIONS.

1. When the sum of the column of units is 157, where do you place the 7, and what do you do with the 15 ?
2. When the sum of the column of tens and of the tens carried from the column of units is 246, what do you do with the 6, and what with the 24 ?

3. In what does the process of the other columns differ from these?

Rem. 1.—Let the class be thoroughly exercised in subtraction, especially in the matter of carrying from one order to another; one example is sufficient to impress it on the mind, as the solution may be repeated as often as is necessary.

Rem. 2.—The teacher can add other questions and larger examples if he think proper.

PRACTICAL EXAMPLES.

1. A farmer has 7 cattle in one field and 8 in another; if he take 4 from the first field and put them in the second, how many will there then be in each field?

2. In one field there are 435 cattle and in another 657; if 320 be taken from the first field and put in the second field, how many cattle are then in each field?

Rem.—In subtraction the larger number is the minuend and the smaller number the subtrahend.

3. A merchant bought dry goods to the amount of 1263 dollars, groceries for 734 dollars, hardware for 231 dollars, and notions for 137 dollars; what is the amount of his purchases?

4. A farmer sold a horse for 175 dollars, cows for 97 dollars, and sheep for 51 dollars; what was the amount of his sales?

5. A gentleman owns five farms; the first is worth 10600 dollars, the second 4970 dollars, the third 5000 dollars, the fourth 6500 dollars, and the fifth 8500 dollars; the amount of his indebtedness is 7984 dollars; if the whole is disposed of at the above rates and the debt paid, what sum will he then have?

ADDITION AND SUBTRACTION. 19

6. A man bought a horse for 125 dollars and sold it for 182 dollars; how much did he gain?

7. A man commenced business with 5000 dollars; the first year his profits were 720 dollars, the second year 500 dollars, the third year 1000 dollars, but the fourth year he lost 2000 dollars; what was then his capital?

8. A man purchased a lot for 900 dollars and erected a house on it at the cost of 3875 dollars for the carpenter's work, 550 dollars for masonry, and the painting cost 869 dollars; he then sold the property for 6000 dollars; did he gain or lose by the transaction, and how much? *Ans.* Lost $194.

9. A man bought a barrel of flour for 8 dollars, three barrels of pork for 35 dollars, salt for 16 dollars, and corn for 300 dollars; he sold the whole so as to gain 20 dollars. How much did he sell it for?

10. A merchant owns property worth 264956 dollars, and owes 89635 dollars; what is the net value of his property?

11. A farmer sold eight cords of wood for 144 dollars; he received in payment cloth valued at 60 dollars, and 48 dollars cash; how much was still owing him?

12. Bought 21693 yards of calico of one merchant, 560 yards of another, and 83946 yards of a third; sold 340 yards to one customer, and 69548 yards to another; how much is still on hand?

13. Sold to one man 3246 acres of land at 6 dollars per acre, to another 4328 acres at 8 dollars per acre, to a third 9546 acres at 5 dollars per acre, and to a fourth 3261 acres at 9 dollars per acre. What was the amount of sales?

MULTIPLICATION.

MULTIPLICATION AND DIVISION TABLE.

1	2	3	4	5	6	7	8	9	10	11	12
2	4	6	8	10	12	14	16	18	20	22	24
3	6	9	12	15	18	21	24	27	30	33	36
4	8	12	16	20	24	28	32	36	40	44	48
5	10	15	20	25	30	35	40	45	50	55	60
6	12	18	24	30	36	42	48	54	60	66	72
7	14	21	28	35	42	49	56	63	70	77	84
8	16	24	32	40	48	56	64	72	80	88	96
9	18	27	36	45	54	63	72	81	90	99	108
10	20	30	40	50	60	70	80	90	100	110	120
11	22	33	44	55	66	77	88	99	110	121	132
12	24	36	48	60	72	84	96	108	120	132	144

As a Multiplication Table, begin with the first line; thus,

Once 1 is 1; twice 1 are 2; three times 1 are 3, etc. Second line, Once 2 are 2; twice 2 are 4; 3 times 2 are 6; 4 times 2 are 8, etc. Third line, Once 3 are 3; twice 3 are 6; 3 times 3 are 9; 4 times 3 are 12, etc. Recite each line similarly.

REM. 4 times 3 are 12, and 3 times 4 are 12; hence, alternating the factors does not change the product.

MULTIPLICATION.

As a Division Table, begin with the first line; thus, 1 is contained in 1, once; in 2, twice; in 3, 3 times; in 4, 4 times, etc. Second line, 2 into $2 = 1$; 2 into $4 = 2$; 2 into $6 = 3$; 2 into $8 = 4$, etc. Third line, 3 into $3 = 1$; 3 into $6 = 2$, etc.

Rem.—As a Multiplication Table, it may also be read by the column, by which the factors are alternated, without changing the product. Any number is multiplied by 10 by adding a zero to it. As a Division Table, the first column has all the divisors, the first line all the quotients, and every number in each line is a dividend, which is always in the same line and the same column with the quotient and divisor. Any number having a zero in the units place is divided by 10 by removing the zero.

THEOREM I.

Any number is multiplied by 10 by annexing a zero to it.

Since the product of any number multiplied by 1 is equal to the number itself, the product of any number multiplied by 2 is double the number, etc.

For, as

$10 \times 1 = 10$, and $10 \times 2 = 20$, and $10 \times 24 = 240$,

and as alternating the factors does not change the product, hence,

$1 \times 10 = 10$, and $2 \times 10 = 20$, and $24 \times 10 = 240$.

∴ Any number is multiplied by 10 by annexing a zero to it.

Cor.—Any number is multiplied by 100 by annexing two zeros to it, and annexing three zeros multiplies it by 1000, etc.

THEOREM II.

The product of any two factors will have as many figures, or one less, than both factors.

1	3	3	4	9	50	500	500
1	3	4	4	9	5	5	50
1	9	12	16	81	250	2500	25000

The products of the smaller figures of units will be but one figure until above 3, when there will be two figures, but never more, as $9 \times 9 = 81$, and every additional figure annexed to each or either factor, whether small or large, will make an increase of one figure and no more; therefore the product of any two factors will have as many figures, or one less than both factors.

Cor. 1.—The product of any two figures cannot be less than one figure, nor more than two.

Cor. 2.—The product of units by units must be units, and when there are two figures, the left-hand figure will be tens. The product of tens by units must be tens, and when there are two figures, the left-hand figure will be hundreds; and if any order be multiplied by units, the right-hand figure of the product will be the same order as the multiplicand, and if there be two figures in the product, the left-hand figure will belong to the next higher order.

Cor. 3.—When the multiplier is tens, the product will be ten times as great as if the multiplier were units; that is, each product will have one zero to the right of it, holding the units place, or the first figure of the product must be placed in the column of tens; when the multi-

MULTIPLICATION.

plier is hundreds, the right-hand figure must be placed in the column of hundreds; and, in general, whatever the order of the multiplier is, the right-hand figure must be in the column of that order.

Cor. 4.—If there be one or more zeros in the multiplier, the product of the next figure will be put back one figure for every zero.

Rem.—In the multiplication, each figure may be regarded as the unit of its order.

PROBLEMS.

1. $10 \times 10 = 100$.

2. $11 \times 11 = 121 = 11 \times (10 + 1) = $ $11 \times 1 = 11$
$11 \times 10 = \underline{110}$
121

3. $12 \times 12 = 144 = 12 \times (10 + 2) = $ $12 \times 2 = 24$
$12 \times 10 = \underline{120}$
144

4. Multiply 432 by $4 = (400 + 30 + 2) \times 4$.

∴ $2 \times 4 = 8$ and 432
$30 \times 4 = 120$ $\underline{4}$
$400 \times 4 = \underline{1600}$ 1728
1728

5. Multiply 432 by $14 = 432 \times (10 + 4)$.

∴ $432 \times 4 = 1728$ or 432
$432 \times 10 = \underline{4320}$ $\underline{14}$
6048 1728
$\underline{432}$
6048

Rem.—The problems should be carefully impressed on the mind before proceeding.

6.

```
    432           432 ×   4 =  1728
    124           432 ×  20 =  8640
   1728           432 × 100 = 43200
    864                        53568
    432
   -----
  53568
```

Cor. 1.—When the multiplicand has several figures and the multiplier one that is only units, the first product of units by units will be units, or units and tens; the units must be placed in the right-hand or units place; if there be tens, it must be reserved and placed in or added to the column of tens; in the next product of tens by units, the right-hand figure will be tens, and must be united with the tens reserved, and placed in the column of tens; the left-hand figure, if there be one, must be treated as the previous one, reserved until the next product is obtained, and united with the right-hand figure; the process is the same in every successive order.

Cor. 2.—When the multiplier also has several figures, the process of each successive multiplier is the same, except that the right-hand figure of each product must be placed in the order of its multiplier. (Cor. 3, Prob. 2, page 22.)

Rem.—A multiplicand may be either an abstract or a concrete number, but a multiplier cannot be concrete, as it cannot refer to things, but merely indicates how many times the multiplicand is to be taken; but the product will be of the same name as the multiplicand; for twice $5 are $10; 3 times 20 yards of cloth are 60 yards of cloth; twice 4 are 8; 3 times 4 are 12, etc.

In computation, it is best to regard all numbers as abstract.

MULTIPLICATION.

(7.)	(8.)	(9.)
36425	26432	26432
324	104	3004
145700	105728	105728
72850	26432	79296
109275	2748928	79401728
11801700		

(10.)	(11.)	(12.)
26432	234	123
50004	123	234
105728	702	492
132160	468	369
1321705728	234	246
	28782	28782

Rem.—The product is not changed by alternating the multiplicand and multiplier.

EXAMPLES.

1. Multiply 54326 by 346.
2. Multiply 23748 by 543.
3. Multiply 46874 by 697.
4. Multiply 36975 by 476.
5. Multiply 236874 by 2134.
6. Multiply 9876325 by 356.
7. Multiply 879654 by 2175.
8. Multiply 986432 by 8704.
9. Multiply 326875 by 3005.
10. Multiply 468753 by 2100.

Examples may be added, or the same repeated, as the student will more readily comprehend by repetition than by different examples.

Rem. 1.—In multiplication, two factors are given to find their product.

Rem. 2.—In division, two numbers also are given to find the third; the one called the dividend corresponds to the product in multiplication, the other given number is called the divisor, and the required number is called the quotient; the two latter correspond to the factors in multiplication.

DIVISION.

PROBLEMS.

When the product of two numbers is 4, and one of the numbers is 2, the other number is also 2; for $2 \times 2 = 4$, and 4 divided by 2, or 4 divided into 2 equal parts, each part is 2, that is, the quotient is 2.

1. $9 \div 3 = 3.$
2. $12 \div 2 = 6.$
3. $12 \div 3 = 4.$
4. $16 \div 4 = 4.$
5. $15 \div 3 = 5.$
6. $15 \div 5 = 3.$

COR. 1.—The product of the divisor and quotient equals the dividend.

COR. 2.—The divisor and quotient may be alternated.

$$\begin{array}{r} 24 \\ 6 \\ \hline 18 \\ 6 \\ \hline 12 \\ 6 \\ \hline 6 \\ 6 \\ \hline 0 \end{array}$$

COR. 3.—Division is the reverse of multiplication and addition, and is similar to subtraction; for, it is separating a number into equal parts, which is the same as subtracting the same number from a larger one; that is, subtracting the divisor from the dividend and then from the remainder, repeating this process until there is no remainder, or until the remainder is less than the divisor. 6 is subtracted 4 times, hence it is contained four times. $24 \div 6 = 4.$

DIVISION. 27

```
     (1.)              (2.)              (3.)
10 ) 100 ( 10    11 ) 121 ( 11     12 ) 144 ( 12.
     10               11                12
     ──               ──                ──
      0               11                24
                      11                24
                      ──                ──

     (4.)                          (5.)
11 ) 121 ( 10 + 1            12 ) 144 ( 10 + 2
     110                          120
     ───                          ───
      11                           24
      11                           24
      ──                           ──
```

6. $48 \div 12 = 4$.   12. $120 \div 10 = 12$.
7. $64 \div 8 = 8$.   13. $130 \div 10 = 13$.
8. $96 \div 12 = 8$.   14. $140 \div 10 = 14$.
9. $12 \times 4 = 48$.   15. $10 \times 12 = 120$.
10. $8 \times 8 = 64$.   16. $10 \times 13 = 130$.
11. $12 \times 8 = 96$.   17. $10 \times 14 = 140$.

Cor. 1.—Adding a zero to the right of a number multiplies the number by 10; taking a zero away from the right of a number divides the number by 10.

Divide 60536 by 4; thus,

```
4 ) 60536 ( 10000          or    4 ) 60536
    40000                            15134
    ─────
    20536 ( 5000
    20000
    ─────
     536 ( 100
     400
     ───
     136 ( 30
     120
     ───
      16 (   4
      16   15134
      ──
```

The divisor 4 is contained once in the unit of the highest order of the dividend, which is one ten-thousand; into the remainder 5000 times, then 100, 30 and lastly 4.

DIVISION.

Rem. 1.—The same result is obtained by short division, by putting the first figure of the quotient under the left-hand figure of the dividend (when it is contained in it), as it is of the same order.

Rem. 2.—If the unit of the divisor is not contained in the first unit of the dividend, then the first figure of the quotient will be of the same order as the second figure of the dividend and should be placed under it.

Divide 60536 by 14; thus,

```
 14 ) 60536 ( 4324        and      214 ) 925336 ( 4324
      56                                  856
      ──                                  ───
      45                                  693
      42                                  642
      ──                                  ───
       33                                  513
       28                                  428
       ──                                  ───
        56                                  856
        56                                  856
       ──                                  ───
```
$4324 \times 14 = 60536.$ $4324 \times 214 = 925336.$

Cor. 1.—Since the product of any two factors will have as many figures or one less than both factors, so in division the number of figures of the divisor and quotient will either be equal to or one greater than that of the dividend.

Cor. 2.—When the divisor is contained in the same number of figures of the dividend as is in the divisor, then the number of figures of the divisor and quotient will be one more than that of the dividend; but when it requires an additional figure of the dividend to contain the divisor, then the number of figures of the divisor and quotient will be equal to that of the dividend.

DIVISION.

PROBLEMS.

1. Divide 9253360 by 2140; thus,

```
214|0 ) 925336|0 ( 4324
       856
       ---
       693    4324
       642    2140
       ---    ----
       513   17296
       428    4324
       ---    ----
       856    8648
       856 9253360
```

(2.)
```
      26432
        104
      -----
     105728
      26432
26432 ) 2748928 ( 104
        26432
        -----
        105728
        105728
```

3. Divide 987654321 by 12300.

```
123|00 ) 9876543|21 ( 80297
         984
         ---
         365
         246
         ---
        1194
        1107
        ----
         873
         861
         ---
        1221, remainder.
```

```
     80297
     12300
     -----
    240891
    160594
     80297
    -------
   987653100
       1221
   ---------
   987654321
```

Cor. 1.—When the dividend is not the exact product of two integral numbers, there will be a remainder, and the dividend is equal to the product of the quotient and divisor plus the remainder.

Cor. 2.—When there are the same number of zeros in dividend and divisor, beginning with the order of units they may be canceled; and when there are zeros in the divisor only, they may be omitted, and also the same number of figures in the dividend, which after the division is performed must be brought down as a part or the whole of the remainder.

DIVISION.

EXAMPLES.

1. Divide 235643 by 123.
2. Divide 345678 by 234.
3. Divide 234567 by 891.
4. Divide 1357916 by 248.
5. Divide 369875432 by 1768.
6. Divide 487698425 by 625.
7. Divide 987654321 by 1234.
8. Divide 876543219 by 2345.
9. Divide 678956732 by 1546.
10. Divide 34567890 by 2564.
11. Divide 786954321 by 176543.
12. Divide 678900432 by 1004000.

The student must not proceed until he is familiar with division.

EXAMPLES.

1. What cost 5 lbs. of sugar at 10 cts. per lb.?
2. At 10 cts. per lb., how many lbs. can be bought for 50 cts.?
3. What cost 10 lbs. of sugar at 10 cts. per lb.?
4. At 10 cts. per lb., how many lbs. can be bought for 100 cts.?
5. What cost 15 lbs. of sugar at 10 cts. per lb.?
6. At 10 cts. per lb., how many lbs. can be bought for 150 cts.?
7. What cost 20 lbs. of sugar at 10 cts. per lb.?
8. At 10 cts. per lb., how many lbs. can be bought for 200 cts.?
9. What cost 25 lbs. of sugar at 10 cts. per lb.?
10. At 10 cts. per lb., how many lbs. can be bought for 250 cts.?

DIVISION.

11. What is the cost of 4 barrels of molasses at 18 dollars a barrel?

12. What is the cost of 9 barrels of sugar at 15 dollars a barrel?

13. What is the cost of 25 pounds of beef at 9 cents a pound?

14. What is the cost of 33 pounds of butter at 21 cents a pound?

15. What is the cost of 147 pounds of cheese at 16 cents a pound?

16. What is the cost of 123 barrels of cider at 4 dollars a barrel?

17. What is the cost of 436 barrels of flour at 8 dollars a barrel?

18. What is the cost of 432 tons of coal at 6 dollars a ton?

19. In an orchard there are 9 rows of trees and 43 trees in each row; how many trees are there in the orchard?

20. Bought 6 pieces of cloth, each containing 42 yards, at 6 dollars a yard; how many yards were there, and what did the whole cost?

21. Bought 9 pieces of cloth, each containing 43 yards, at 5 dollars a yard, and 25 barrels of flour at 6 dollars a barrel; what did the whole cost?

22. If 4 barrels of flour cost 32 dollars, what is that per barrel?

23. A man paid 320 dollars for two horses; what was the price of each?

24. A man owning 480 acres of land, wishes to divide it equally among his 4 sons: how many acres will each get?

25. A farm of 120 acres is laid off into six fields of equal size; how many acres in each?

26. A man is 45 years old, and he is three times as old as his son. How old is the son?

27. A farmer sold 325 bushels of wheat for 650 dollars. What was the rate per bushel?

28. In an orchard there are 625 trees, and there are 25 trees in a row. How many rows are there?

29. In a garden there are 100 heads of cabbage in ten rows, and an equal number in each row. How many heads in each row?

30. A farmer sowed five bushels of clover-seed on forty acres of land. How many acres to a bushel of seed?

31. A man dying, bequeathed his estate of 580,640 dollars, as follows: 60,000 dollars to each of two daughters, 125,000 dollars to a son, 22,000 dollars to each of his four sisters, and the remainder to his grandson. What was the grandson's share?

32. Four persons enter into partnership; the first invests 8564 dollars, the second 500 dollars more than the first, the third as much as the first and second, and the fourth 645 dollars less than the third. What was the whole amount invested?

33. Bought silk for 1600 dollars, lace for 960 dollars, shoes for 356 dollars, a shawl for 30 dollars, and some ribbon for 5 dollars. Paid at one time 500 dollars, at another 685 dollars, and at a third 800 dollars. How much is still owing?

34. A man bought a carriage and two horses for 903 dollars; the horses were valued at 450 dollars. What

was the value of the carriage, and how much more was the carriage than the horses?

35. Bought molasses for 2634 dollars, sugar for 965 dollars, and coffee for 1589 dollars; sold the molasses at a gain of 500 dollars, the sugar at cost, and lost 100 dollars on the coffee. Did I gain or lose, and how much?

36. Washington was born in 1732 and died in 1799; how old was he when he died?

37. The pyramids of Egypt were built 337 years before the founding of Carthage; Carthage was founded 49 years before the destruction of Troy; Troy was destroyed 431 years before Rome was founded; Carthage was destroyed 607 years after the founding of Rome and 146 years before the Christian era. How many years before Christ were the pyramids built?

38. What is the cost of 5 pounds of sugar, at 12 cents a pound?

39. What is the cost of 15 pounds of sugar, at 11 cents a pound?

40. What is the cost of 335 pounds of sugar, at 9 cents a pound?

41. What is the cost of 23 tons of hay, at 14 dollars a ton?

42. What is the cost of 435 acres of land, at 43 dollars per acre?

43. What is the cost of 753 bushels of oats, at 35 cents per bushel?

44. If 5 lbs. of sugar cost 60 cents, what is it per lb.?

45. If 15 lbs. of sugar cost 165 cents, what is it per lb.?

46. If sugar at 9 cents per lb. cost 3015 cents, how many pounds were there?

DIVISION.

47. How many tons of hay can be bought for 322 dollars, at 14 dollars per ton?

48. How many acres of land will 18705 dollars buy, at 43 dollars per acre?

49. If 753 bushels of oats cost 26355 cents, what is the price of a bushel?

50. If 5 lbs. of sugar cost 60 cts., what will 12 lbs. cost?

51. If 15 lbs. of sugar cost 165 cts., what will 80 lbs. cost?

52. If 335 lbs. of sugar cost 3015 cts., what will 36 lbs. cost?

53. If 23 tons of hay cost 322 dollars, what will 69 tons cost?

54. If 435 acres of land cost 18705 dollars, what will 87 acres cost?

55. If 753 bushels of oats cost 26355 cents, what will 251 bushels cost?

56. If 100 acres of land cost 5625 dollars, what will 1000 acres cost?

57. If 1000 acres cost 78560 dollars, what will 100 acres cost?

Rem. 1.—If 1 pound of sugar cost 12 cents, 5 lbs. will cost 5 times 12 cents = 60 cents, or 5 lbs. sugar at 1 cent a pound will cost 5 cts., and at 12 cts. per lb., 12 times 5 = 60 cts.; if 5 lbs. of sugar cost 60 cts., and it be divided into five equal parts, 1 lb. will cost 12 cts., and 12 lbs. will cost 12 times 12 cts. = 144 cts.

Rem. 2.—In the 53d example, 69 is 3 times 23, and will cost 3 times 322 dollars; in the 54th, 435 is 5 times 87, and must be divided into 5 equal parts, that is, divided by 5; and in the 55th, 753 is 3 times 251, and must be divided by 3. The 56th example is solved by adding one zero to the number of dollars, and the 57th by taking away a zero.

DIVISION.

PROBLEMS.

1. The sum of two numbers, and one of the numbers given, to find the other number.

The difference between the given number and the sum will be the other number.

2. The difference between two numbers and the smaller number given, to find the larger one.

The sum of the difference and the smaller number is the larger one.

3. The difference of two numbers and the larger one given, to find the smaller one.

Subtract the difference from the larger number; the remainder will be the smaller one.

4. The sum and difference of two numbers given, to find the numbers.

Subtract the difference from the sum, and divide the remainder by 2; the quotient will be the smaller one, to which add the difference, and the sum will be the larger one.

5. The product of two numbers and one of the numbers given, to find the other.

Divide the product by the given number; the quotient will be the other number.

6. The product of three numbers, and two of the numbers given, to find the third.

Divide the product of the three by the product of the two.

FACTORING.

The **Product** of two or more numbers is found by multiplication, and the factors of that product are restored by division.

Any number that is the product of two or more numbers is called a *Composite Number;* as, 12 is the product of 4 and 3, and 4 is the product of 2 and 2; hence, the factors of 12 are 2, 2, and 3; these factors cannot be reduced, they are therefore called *Prime Factors;* as any number is called Prime which is not formed by other factors than itself and unity.

The Prime Factors are readily found; thus, take the first fifty numbers,

1, 2, 3, ~~4~~, 5, ~~6~~, 7, ~~8~~, ~~9~~, ~~10~~,
11, ~~12~~, 13, ~~14~~, ~~15~~, ~~16~~, 17, ~~18~~, 19, ~~20~~,
~~21~~, ~~22~~, 23, ~~24~~, ~~25~~, ~~26~~, ~~27~~, ~~28~~, 29, ~~30~~,
31, ~~32~~, ~~33~~, ~~34~~, ~~35~~, ~~36~~, 37, ~~38~~, ~~39~~, ~~40~~,
41, ~~42~~, 43, ~~44~~, ~~45~~, ~~46~~, 47, ~~48~~, ~~49~~, ~~50~~.

Every even number after 2 is composite, as it is divisible by 2, "strike these;" every third number after 3 is divisible by 3, strike these; every fifth number after 5; every seventh number after 7; the 9's are canceled by 3; every 11th after 11, etc.; in the above there were none after the 7's; when more numbers are taken higher

FACTORING.   37

numbers will be required. The prime numbers in the first fifty are sixteen, viz.,

 1, 2, 3, 5, 7, 11, 13, 17,
 19, 23, 29, 31, 37, 41, 43, 47.

The following Corollaries enable us readily to discover the Prime Factors of numbers:

1. Every even number is divisible by 2.

2. Every number, whose last two figures express a number which is a multiple of 4, is divisible by 4; for if the number expressed by these two figures is subtracted from the whole number, the remainder will be a certain number of hundreds which are divisible by 4.

3. Every number ending in 5 is divisible by 5.

4. Every number ending with zero is divisible by 10, consequently by 2 and 5.

5. Every number is divisible by 3, when the sum of its figures taken as units is divisible by 3; for if from 1000 one be subtracted, the remainder is divisible by 9; if from 100 one be subtracted, the remainder is divisible by 9; so also if one is subtracted from 10; hence, if from 2000 two be subtracted, the remainder is divisible by 9; so also 2 from 200, or 2 from 20; therefore, in dividing by nine, any number of thousands, hundreds or tens, the remainder will always be the unit of the thousands, hundreds, and tens; consequently if the sum of all these remainders as units, and also of the units of the given number, equals 9 or any number of 9's, then the whole number is divisible by 9, and 9 is divisible by 3.

FACTORING.

Resolve the following members into their Prime Factors:

1. $30 = 2, 3, 5.$
2. $32 = 2, 2, 2, 2, 2.$
3. $33 = 3, 11.$
4. $34 = 2, 17.$
5. $35 = 5, 7.$
6. $36 = 2, 2, 3, 3.$
7. $38 = 2, 19.$
8. $39 = 3, 13.$
9. $40 = 2, 2, 2, 5.$
10. $42 = 2, 3, 7.$
11. $44 = 2, 2, 11.$
12. $45 = 3, 3, 5.$
13. $46 = 2, 23.$
14. $48 = 2, 2, 2, 2, 3.$
15. $49 = 7, 7.$
16. $50 = 2, 5, 5.$

17. 51.
18. 52.
19. 54.
20. 55.
21. 56.
22. 57.
23. 58.
24. 60.
25. 62.
26. 63.
27. 64.
28. 65.
29. 68.
30. 70.
31. 72.
32. 74.
33. 75.
34. 76.
35. 77.
36. 78.
37. 80.
38. 81.
39. 82.
40. 84.
41. 85.
42. 86.
43. 87.
44. 88.
45. 90.
46. 91.
47. 92.
48. 94.
49. 105.
50. 210.
51. 336.
52. 540.

COR.—When a number is resolved into its prime factors, the original number is divisible by all these prime factors, and by all the quotients arising from these factors as divisors of the original number; and when a factor occurs more than once, by the products of the like factors.

DEF.—One number is called a *Multiple* of another number, when it is exactly divisible by that other number.

FACTORING. 39

Find all the numbers of which the following numbers are multiples:

	FACTORS.	DIVISORS.
45.	3, 3, 5.	3, 5, 9, 15.
60.	2, 2, 3, 5.	2, 3, 5, 4, 6, 10, 12, 15, 20, 30.
70.	2, 5, 7.	2, 5, 7, 10, 14, 35.
72.	2, 2, 2, 3, 3.	2, 3, 6, 8, 9, 12, 18, 36.
75.	3, 5, 5.	3, 5, 15, 25.
80.	2, 2, 2, 2, 5.	2, 5, 4, 8, 10, 16, 20, 40.
100.	2, 2, 5, 5.	2, 4, 5, 10, 20, 25, 50.

PROBLEMS.

1. Find the least common multiple of 9 and 15.

$$9 \quad 15$$
$$3, 3. \quad 3, 5.$$
$$3 \times 3 \times 5 = 45.$$

It is evident that any number which contains all the prime factors of each number is a common multiple of the given numbers; and the least common multiple must contain these factors and no others; and if the same factor occur several times in any number, it must occur just as often in the multiple.

2. Find the least common multiple of 6 and 12.

$$6 \quad 12$$

As twelve is a multiple of 6, it contains all the prime factors of 6; hence, when any given number is a multiple of another given number, the later may be canceled.

3. Find the least common multiple of 6, 8, 10, 12, 15.

$$\begin{array}{r|rrrrr} 2 & \cancel{6} & 8 & 10 & 12 & 15 \\ 2 & & 4 & \cancel{5} & 6 & 15 \\ \hline & & & 2 & \cancel{3} & 15 \end{array}$$
$$2 \times 2 \times 2 \times 15 = 120$$

As 12 is a multiple of 6, cancel the 6; and as 2 is a common factor of two or more numbers, divide the 2 into them,

40 FACTORING.

reserving the divisor, the quotients, and the numbers not divided; as 15 is a multiple of 5, cancel 5; and as 2 is now a common factor of two of the quotients, divide it into them, reserving as before; and as 15 is now a multiple of 3, cancel 3; the product of all the divisors and of the remaining quotients and original numbers, if there be any, will be the least common multiple.

EXAMPLES.

1. Find the least common multiple of 6 and 15.
Ans. 30.
2. Find the least common multiple of 6, 15, and 42.
Ans. 210.
3. Find the least common multiple of 10, 12 and 14.
Ans. 420.
4. Find the least common multiple of 4, 6, and 8.
Ans. 24.
5. Find the least common multiple of 6, 8, and 10.
Ans. 120.
6. Find the least common multiple of 12, 18, 27, and 36. *Ans.* 108.
7. Find the least common multiple of 10, 15, 25, and 40. *Ans.* 600.
8. Find the least common multiple of 14, 18, 21, and 28. *Ans.* 252.
9. Find the least common multiple of 15, 25, 36, and 48. *Ans.* 3600.
10. Find the least common multiple of 25, 45, 70, and 90. *Ans.* 3150.
11. Find the least common multiple of 4, 8, 16, 32, and 64. *Ans.* 64.
12. Find the least common multiple of 3, 7, 11, and 13. *Ans.* 3003.

GREATEST COMMON DIVISOR.

The *Greatest Common Divisor* of two or more numbers, is the highest number that will exactly divide the numbers.

Cor.—The greatest common divisor of two or more numbers must be the factor or the product of all the factors which are common to the given numbers.

PROBLEMS.

Find the greatest common divisor of the following numbers:

1. Of 6 and 14.
     2, 3. 2, 7.
The only factor common is $2 =$ G. C. D.

2. Of 12 and 18.
       $12 = 2, 2, 3,$ and $18 = 2, 3, 3$.
One 2 and one 3 are common: therefore,
       $2 \times 3 = 6 =$ G. C. D.

3. Of 42 and 70.
       $42 = 2, 3, 7,$ and $70 = 2, 5, 7;$
     $\therefore$ $2 \times 7 = 14 =$ G. C. D.

```
2 | 84   126   210
3 | 42    63   105
7 | 14    21    35
       2     3     5
```

4. Of 84, 126, and 210.
 By this method the common factors are readily found.
    $2 \times 3 \times 7 = 42.$

The process of the last example is the shortest for finding the greatest common divisor of small numbers; but when the numbers are large and the common factors not so readily found, the following method is generally adopted.

5. Find the G. C. D. of 84 and 147.

```
84 ) 147 ( 1
     84
     ──
     63 ) 84 ( 1
          63
          ──
          21 ) 63 ( 3
               63
               ──
```

Divide the smaller number into the larger, and the remainder into the last divisor; and again the remainder into the last divisor, until there is no remainder; the last divisor is the G. C. D. of the two numbers.

ANALYSIS.—As each number is a multiple of the G. C. D., so the difference of the numbers is also a multiple of the G. C. D.; hence in every case the dividend and divisor are multiples of the G. C. D., and whenever the divisor is contained in the dividend, that divisor is the G. C. D.

6. Find the G. C. D. of 323 and 425. G. C. D = 17.
7. Find the G. C. D. of 2310 and 4626. G. C. D. = 6.

CANCELLATION.

THEOREM.

The dividend contains all and exactly the same factors as the divisor and quotient.

Any composite number is the product of all its prime factors, and may be resolved into them. The product of any two integral numbers is a composite number and must contain all the factors of both numbers; and as a dividend is the product of its divisor and quotient, it must contain the same factors as its divisor and quotient.

Cor. 1.—The same is true if one or both divisor and quotient be fractional; for when reduced to a common denominator, their numerators may be regarded as integral.

Cor. 2.—Every factor of the divisor will cancel the same factor in the dividend.

Cor. 3.—The factors which are not canceled by those of the divisor will be the factors of the quotient.

Cor. 4.—Canceling a factor in the dividend divides the quotient by the same factor.

Cor. 5.—Canceling a factor in the divisor multiplies the quotient by the same factor.

PROBLEMS.

1. Divide 648 by 36.

$$\frac{648}{36} = \frac{\not{2}, \not{2}, 2, 3, 3, \not{3}, \not{3}}{\not{2}, \not{2}, \not{3}, \not{3}} = 18, \textit{Ans.}$$

2. Divide 625 by 125.

$$\frac{625}{125} = \frac{\not{5}, \not{5}, \not{5}, 5.}{\not{5}, \not{5}, \not{5}.} = 5, \textit{Ans.}$$

3. Divide 500 by 100.

$$\frac{5\not{0}\not{0}}{1\not{0}\not{0}} = 5, \textit{Ans.}$$

4. A man bought 30 yards of cloth at $5 a yard; he then exchanged it for other cloth at $3 a yard. How many yards of the latter did he get?

PRACTICAL EXAMPLES.

1. A man bought 30 cows at 25 dollars each; he then exchanged the cows for horses at 50 dollars each; how many horses did he get? *Ans.* 15 horses.

2. A farmer sold 1500 bushels of wheat at 125 cents per bushel, and received in return barley at 75 cents per bushel. How many bushels of barley did he get? *Ans.* 2500 bu. barley.

3. How many acres of land, at 25 dollars per acre, can be obtained for 5 houses and lots at 750 dollars each? *Ans.* 150 acres.

4. How many yards of flannel, three-quarters of a yard wide, will line a coat made of 3 yards of cloth six-quarters wide? *Ans.* 6 yards.

5. Sold 320 acres of land at 60 dollars per acre, and invested the proceeds in other land at 40 dollars per acre; how many acres did I get? *Ans.* 480 acres.

6. Exchanged 432 pieces of cloth at 18 dollars each, for linen at 6 dollars a piece. How many pieces of linen did I get? *Ans.* 1296 pieces.

7. Sold a farm of 477 acres at 48 dollars per acre, and invested the returns of sale in another farm at $36 per acre. How many acres did I buy? *Ans.* 636 acres.

8. Sold 15 horses at 732 dollars each, and bought sheep for the proceeds at 4 dollars each. How many sheep did I buy? *Ans.* 2745.

9. Multiply the following numbers: 12, 15, 27, 28, 32, and divide the product by 2, 3, 4, 5, 6, 7, 8, and 9. *Ans.* 12.

FRACTIONS.

Def. 1.—If a unit or any other number is divided into equal parts, one or more of these parts is a fraction of the whole, and all the parts constitute the whole. If a unit is divided into two equal parts, each part is called one-half, and is written $\frac{1}{2}$; and the two halves constitute the whole; thus, $\frac{2}{2} = 1$. If a unit is divided into three equal parts, each part is one-third ($\frac{1}{3}$); two of the parts, $\frac{2}{3}$; and the three thirds constitute the whole; thus, $\frac{3}{3} = 1$. If 5 is divided into two equal parts, each part is five-halves ($\frac{5}{2}$); two of the parts, $\frac{10}{2} = 5$, etc.; if 6 is divided into three equal parts, each part is $\frac{6}{3} = 2$; and two-thirds of 6 is 4, etc. Fourths, fifths, sixths, etc., are similarly constructed.

2. When a unit is divided into equal parts, any number of the parts less than the whole, expressed fractionally, is called a *Proper Fraction;* as, $\frac{1}{2}$, $\frac{2}{3}$, $\frac{3}{4}$, $\frac{9}{10}$, etc. The quotient is the same in division when the dividend is less than the divisor; as, $\frac{9}{13}$, $\frac{14}{19}$, etc.; but when the divisor is less than the dividend, the quotient is called an *Improper Fraction;* as, $\frac{5}{2}$, $\frac{10}{3}$, $\frac{14}{4}$, etc.

3. When the division indicated by an Improper Fraction is performed, and the divisor is not contained an exact number of times in the dividend, the quotient is partly integral and partly fractional, and is termed a

Mixed Number; thus, $\frac{5}{4} = 1\frac{1}{4}$; $\frac{50}{6} = 8\frac{2}{6}$; and $\frac{31}{3} = 10\frac{1}{3}$.

Cor.—The denominator expresses the number of parts into which a unit or any other number is divided, and the numerator expresses the number of parts of a unit taken, or the number divided.

Exemplification.—If each half is divided into two equal parts, the whole number of parts is four, and the one-half has made two of those parts; hence, $\frac{1}{2} = \frac{2}{4}$; if each half is divided into three equal parts, the whole number of parts is six, and $\frac{1}{2} = \frac{3}{6}$; $\therefore \frac{1}{2} = \frac{2}{4} = \frac{3}{6} = \frac{4}{8} = \frac{5}{10}$, etc., and $\frac{1}{3} = \frac{2}{6} = \frac{3}{9} = \frac{4}{12} = \frac{5}{15}$, etc.; hence, if both terms of a fraction are multiplied by the same number, the value of the fraction is not changed, and by this principle fractions are reduced to a Common Denominator.

PROBLEMS.

1. Reduce $\frac{1}{2}$ and $\frac{1}{3}$ to a common denominator.
$$\tfrac{1\times 3}{2\times 3}=\tfrac{3}{6}, \text{ and } \tfrac{1\times 2}{3\times 2}=\tfrac{2}{6}.$$

2. Reduce $\frac{1}{2}$, $\frac{1}{3}$, and $\frac{1}{4}$ to a common denominator.
$$\tfrac{1\times 6}{2\times 6}=\tfrac{6}{12},\ \tfrac{1\times 4}{3\times 4}=\tfrac{4}{12},\text{ and } \tfrac{1\times 3}{4\times 3}=\tfrac{3}{12}.$$

3. Reduce $\frac{4}{5}$ and $\frac{6}{7}$ to a common denominator.
$$\tfrac{4\times 7}{5\times 7}=\tfrac{28}{35},\text{ and } \tfrac{6\times 5}{7\times 5}=\tfrac{30}{35}.$$

Cor. 1.—The least common multiple of all the denominators is the least common denominator.

Cor. 2.—Fractions are reduced to a common denominator thus: Multiply both terms of each fraction by the quotient obtained by dividing its denominator into the least common denominator; or, when all the denom-

FRACTIONS.

inators are prime to each other, multiply both terms of each fraction by all the other denominators.

Cor. 3.—An integral number is reduced to a fraction by multiplying it by the denominator of the fraction.

EXAMPLES.

1. Reduce $\frac{2}{3}$, $\frac{3}{4}$, and $\frac{5}{6}$ to a common denominator.
     Ans. $\frac{48}{72}$, $\frac{54}{72}$, and $\frac{60}{72}$.
2. Reduce $\frac{1}{4}$, $\frac{5}{6}$, and $\frac{3}{8}$ to a common denominator.
     Ans. $\frac{6}{24}$, $\frac{20}{24}$, and $\frac{9}{24}$.
3. Reduce $\frac{4}{5}$, $\frac{7}{8}$, and $\frac{8}{9}$ to a common denominator.
     Ans. $\frac{288}{360}$, $\frac{315}{360}$, and $\frac{320}{360}$.
4. Reduce $\frac{2}{3}$, $\frac{5}{6}$, $\frac{7}{8}$, and $\frac{9}{11}$ to a common denominator.
     Ans. $\frac{2072}{3465}$, $\frac{1414}{3465}$, $\frac{2310}{3465}$, and $\frac{945}{3465}$.
5. Reduce $\frac{1}{4}$, $\frac{6}{25}$, $\frac{7}{100}$, and $\frac{3}{1000}$ to a common denominator.
     Ans. $\frac{250}{1000}$, $\frac{240}{1000}$, $\frac{70}{1000}$, and $\frac{3}{1000}$.

EXEMPLIFICATION.—Since $\frac{1}{2} = \frac{2}{4} = \frac{3}{6} = \frac{4}{8} = \frac{5}{10}$, etc., and $\frac{1}{3} = \frac{2}{6} = \frac{3}{9} = \frac{4}{12} = \frac{5}{15}$, etc., and $\frac{2+2}{4+2}=\frac{1}{2}$, $\frac{3+3}{6+3}=\frac{1}{2}$, $\frac{4+4}{8+4}=\frac{1}{2}$, etc., and $\frac{2+2}{6+2}=\frac{1}{3}$, $\frac{3+3}{9+3}=\frac{1}{3}$, $\frac{4+4}{12+4}=\frac{1}{3}$, etc.; hence, if both terms of a fraction are divided by the same number, the value of the fraction is not changed.

Cor.—When both terms of a fraction have a common factor, it may be canceled, and the fraction is thereby reduced to its lowest terms.

Rem.—The common factor may be either a prime or a composite number, and it is the greatest common divisor of the terms.

EXAMPLES.

1. Reduce $\frac{9}{15}$ to its lowest terms.   *Ans.* $\frac{9+3}{15+3}=\frac{3}{5}$.
2. Reduce $\frac{14}{21}$ to its lowest terms.   *Ans.* $\frac{2}{3}$.
3. Reduce $\frac{143}{1003}$ to its lowest terms.   *Ans.* $\frac{13}{131}$.

THEOREM.

The least common multiple of two or more fractions has for its numerator the least common multiple of all the numerators, and for its denominator the greatest common divisor of all the denominators.

The least common multiple of all the numerators will evidently be a common multiple of all the fractions; for each numerator regarded as an integer is a multiple of its denominator regarded as a fraction; and when the denominators are prime to each other, this common multiple will be the least common multiple of the fractions; but when all the denominators have a common factor, then the value of each fraction is reduced by it, and the common multiple may be reduced by the same; hence, the greatest common divisor of all the denominators is the denominator of the least common multiple.

EXAMPLES.

1. Find the L. C. M. of $\frac{1}{2}$ and $\frac{1}{3}$. Evidently 1.
2. Find the L. C. M. of $\frac{1}{4}$ and $\frac{1}{6}$. Evidently $\frac{1}{2}$.

THEOREM.

The greatest common divisor of two or more fractions has for its numerator the greatest common divisor of all the numerators, and for its denominator the least common multiple of all the denominators.

The greatest common divisor of all the numerators is evidently the numerator of the greatest common divisor of the fractions; and as multiplying the denominators divides the fraction and the less the denominator the greater the value of the fraction, hence the least common multiple of all the denominators is the denominator of the greatest common divisor of the fractions.

EXAMPLES.

1. Find the greatest common divisor of $\frac{1}{2}$ and $\frac{1}{3}$.

   Evidently $\frac{1}{6}$.

2. Find the greatest common divisor of $\frac{4}{5}$ and $\frac{2}{3}$.

   Evidently $\frac{2}{15}$.

ADDITION AND SUBTRACTION OF FRACTIONS.

EXAMPLES.

1. Add and subtract $\frac{3}{4}$ and $\frac{1}{2}$.   4 is L. C. D.
2. Add and subtract $\frac{2}{3}$ and $\frac{1}{2}$.   6 is L. C. D.
   Ans. Sum, $\frac{7}{6}$; difference, $\frac{1}{6}$.
3. Add and subtract $\frac{3}{4}$ and $\frac{2}{5}$.   20 is L. C. D.
   Ans. Sum, $\frac{23}{20}$; difference, $\frac{7}{20}$.
4. Add and subtract $\frac{5}{6}$ and $\frac{2}{5}$.   30 is L. C. D.
   Ans. Sum, $\frac{37}{30}$; difference, $\frac{13}{30}$.
5. Add and subtract $\frac{5}{6}$ and $\frac{4}{7}$.   42 is L. C. D.
   Ans. Sum, $\frac{59}{42}$; difference, $\frac{11}{42}$.

Cor.—When the denominators have no common factor, then their product is the L. C. D., and each numerator is multiplied by all the denominators except its own.

6. Add $\frac{2}{3}$ and $\frac{3}{4}$.   L. C. D., 12.
$$\frac{2 \times 4}{3 \times 4} = \frac{8}{12}. \quad \frac{3 \times 3}{4 \times 3} = \frac{9}{12}.$$
$8 + 9 = 17.$ Sum $= \frac{17}{12} = 1\frac{5}{12}.$

7. Add $\frac{4}{5}$, $\frac{2}{3}$, and $\frac{4}{5}$.   Sum $= \frac{133}{60} = 2\frac{13}{60}.$

As no two numbers have a common factor, the L. C. D. is the product of all the denominators; and then as each denominator is multiplied by the other two denominators, so each numerator must be multiplied by the product of all the denominators except its own.

8. Add $\frac{4}{5}$, $\frac{5}{8}$, and $\frac{7}{12}$.

$$\begin{array}{r|ccc} 4 & 5 & 8 & 12 \\ \hline & 5 & 2 & 3 \end{array}$$

$4 \times 5 \times 2 \times 3 = 120.$

$\frac{120}{5} = 24.$ $\frac{120}{8} = 15.$ $\frac{120}{12} = 10.$

∴ 24, 15, and 10 are the multipliers of the fractions.

ADDITION OF FRACTIONS.

EXAMPLES.

1. Add $\frac{1}{2}$ and $\frac{1}{4}$.

$\frac{1}{2} = \frac{2}{4}$, and $\frac{2}{4} + \frac{1}{4} = \frac{3}{4}$ = Sum.

2. Add $\frac{1}{2}$ and $\frac{1}{5}$. 10 is the least common denominator. $\frac{1 \times 5}{2 \times 5} = \frac{5}{10}$ and $\frac{1 \times 2}{5 \times 2} = \frac{2}{10}$; and $\frac{5}{10} + \frac{2}{10} = \frac{7}{10}$ = Sum.

3. Add $\frac{1}{2}$, $\frac{1}{3}$ and $\frac{1}{4}$. 12 = least common denominator.

$\frac{1}{2} \times \frac{6}{6} = \frac{6}{12}$, $\frac{1 \times 4}{3 \times 4} = \frac{4}{12}$ and $\frac{1 \times 3}{4 \times 3} = \frac{3}{12}$, and $\frac{6}{12} + \frac{4}{12} + \frac{3}{12} = \frac{13}{12} = 1\frac{1}{12}$ = Sum.

4. Add $\frac{1}{3}$, $\frac{1}{4}$, $\frac{3}{5}$ and $\frac{7}{3}$.

$$3 \;)\; \underline{3 \quad 4 \quad 5 \quad 9}$$
$$ 1 \quad 4 \quad 5 \quad 3 \quad \therefore \; 3 \times 4 \times 5 \times 3 = 180.$$

$$3 \;)\; \underline{180} \quad 4 \;)\; \underline{180} \quad 5 \;)\; \underline{180} \quad 9 \;)\; \underline{180}$$
$$ 60   45   36   20$$

Rem.—The terms of the 1st fraction must be multiplied by 60, the 2d by 45, the 3d by 36, and the 4th by 20.

Cor.—Fractions are reduced to a common denominator by multiplying both terms of the fraction by the quotient, obtained by dividing the common denominator by the denominator of each given fraction.

5. Add $\frac{5}{7}$, $\frac{7}{9}$ and $\frac{9}{11}$.

$5 \times 9 \times 11 = 495$
$7 \times 7 \times 11 = 539$
$9 \times 7 \times 9 = \underline{567}$
1601
$7 \times 9 \times 11 = \overline{693} = 693\;)\;1601\;(\;2\frac{215}{693}$ = Sum.
$\underline{1386}$
215
$\overline{693}$

FRACTIONS.

REM.—When the denominators are prime to each other, as no two have a common factor, the least common denominator is the product of all the denominators, and each numerator is multiplied by all the denominators except its own.

6. Add $\frac{2}{3}, \frac{1}{7}$ and $\frac{1}{2}$.

$$\frac{2}{3}\frac{1}{7}\frac{1}{2} = 1\frac{1}{4}\frac{1}{2}\frac{1}{2}.$$

7. Add $\frac{1}{8}, \frac{1}{5}$ and $\frac{3}{12}$. Here 120 is the common denominator.

$$\frac{1}{1}\frac{2}{2}\frac{3}{0} = 1\frac{1}{15}.$$

8. Add $\frac{5}{12}, \frac{7}{15}, \frac{1}{4}$ and $\frac{1}{6}$. $1\frac{2}{3} = $ Sum.

SUBTRACTION OF FRACTIONS.
EXAMPLES.

1. Subtract $\frac{4}{15}$ from $\frac{11}{12}$.
   C. D. = 60   Difference = $\frac{1}{2}$.
2. Subtract $\frac{7}{11}$ from $\frac{15}{16}$.
   C. D. = 176.   Difference = $\frac{2}{1}\frac{1}{76}$.
3. Subtract $\frac{4}{7}$ from $\frac{11}{12}$.
   C. D. = 84.   Difference = $\frac{3}{7}$.
4. Subtract $\frac{21}{57}$ from $\frac{4}{4}$.
   C. D. = 2793.   Difference = $\frac{5}{27}\frac{4}{9}\frac{4}{3}$.
5. Subtract $\frac{11}{16}$ from $\frac{3}{9}$.   Difference = $\frac{4}{14}\frac{3}{4}$.

REM.—The above examples should be repeated, or similar ones given, until the class is familiar with addition and subtraction of fractions.

6. Subtract $\frac{4}{9}$ from $\frac{8}{11}$.   · Difference = $\frac{1}{2}$.
7. Subtract $\frac{4}{15}$ from $\frac{11}{12}$.   Difference = $\frac{79}{105}$.
8. Subtract $5\frac{1}{2}$ from $8\frac{4}{5}$.   Difference = $3\frac{1}{4}$.
9. Subtract $3\frac{4}{5}$ from $5\frac{1}{4}$.   Difference = $1\frac{7}{11}$.

REM.—Reduce Ex. 8 and 9 to improper fractions.

MULTIPLICATION OF FRACTIONS.

THEOREM.

The product of two proper fractions is less than either fraction.

For, if a number is multiplied by one, the product is the same as the number multiplied. If the multiplier is greater than one, the product is greater than the number; and if the multiplier is less than one, the product is less than the number.

In the multiplication of two proper fractions, each factor is less than one; hence the product is less than either fraction.

PROBLEMS.

1. Multiply $\frac{2}{3}$ by 1.   *Ans.* $\frac{2}{3} \times 1 = \frac{2}{3}$.
2. Multiply $\frac{2}{3}$ by 2.   *Ans.* $\frac{2}{3} \times 2 = \frac{4}{3}$.
3. Multiply $\frac{2}{3}$ by $\frac{1}{2}$.   *Ans.* $\frac{2}{3} \times \frac{1}{2} = \frac{1}{3}$.

It is evident that $\frac{2}{3}$ multiplied by $1 = \frac{2}{3}$, that $\frac{2}{3} \times 2$ is $\frac{4}{3}$, and $\frac{2}{3} \times \frac{1}{2}$ or $\frac{1}{2}$ time $\frac{2}{3}$ is $\frac{1}{3}$; and, as alternating the factors does not change the product, therefore, $1 \times \frac{2}{3} = \frac{2}{3}$, $2 \times \frac{2}{3} = \frac{4}{3}$, and $\frac{1}{2} \times \frac{2}{3} = \frac{1}{3}$.

4. Multiply $\frac{3}{4}$ by $\frac{2}{3}$.   *Ans.* $\frac{3}{4} \times \frac{2}{3} = \frac{2}{4}$.

Cor. 1.—In multiplying by a fraction the numerator is a multiplier and the denominator a divisor.

Cor. 2.—In the multiplication of fractions, the product of all the numerators is the numerator of the product; and the product of all the denominators is the denominator of the product.

FRACTIONS. 53

Cor. 3.—The product of two improper fractions is greater than either fraction.

EXAMPLES.

1. Multiply $\frac{1}{2}, \frac{2}{3}, \frac{3}{4}, \frac{4}{5}, \frac{5}{6}, \frac{6}{7}, \frac{7}{8}, \frac{8}{9}, \frac{9}{10}, \frac{10}{11}, \frac{11}{12}$.

$\frac{1}{2} \times \frac{2}{3} \times \frac{3}{4} \times \frac{4}{5} \times \frac{5}{6} \times \frac{6}{7} \times \frac{7}{8} \times \frac{8}{9} \times \frac{9}{10} \times \frac{10}{11} \times \frac{11}{12} = \frac{1}{12}$.

By analysis, $\frac{1}{2}$ of $\frac{2}{3} = \frac{1}{3}$, $\frac{1}{3}$ of $\frac{3}{4} = \frac{1}{4}$, $\frac{1}{4}$ of $\frac{4}{5} = \frac{1}{5}$, $\frac{1}{5}$ of $\frac{5}{6} = \frac{1}{6}$, $\frac{1}{6}$ of $\frac{6}{7} = \frac{1}{7}$, $\frac{1}{7}$ of $\frac{7}{8} = \frac{1}{8}$, $\frac{1}{8}$ of $\frac{8}{9} = \frac{1}{9}$, $\frac{1}{9}$ of $\frac{9}{10} = \frac{1}{10}$, $\frac{1}{10}$ of $\frac{10}{11} = \frac{1}{11}$, $\frac{1}{11}$ of $\frac{11}{12} = \frac{1}{12}$.

2. Multiply $\frac{7}{3}$, $\frac{10}{11}$, and $\frac{2}{3}$; thus,

$$\overset{2}{\underset{3}{\frac{7}{3}}} \times \frac{10}{11} \times \frac{2}{3} = \frac{10}{11}, \text{ product.}$$

3. Multiply $\frac{9}{10}$ and $\frac{3}{4}$.

$\frac{9}{10} \times \frac{3}{4} = \frac{27}{40}$, product.

4. Multiply $35\frac{3}{4}$ by 9.

$$\begin{array}{rl} \frac{3}{4} \times 9 = \frac{27}{4} = & 6\frac{3}{4} \\ 35 \times 9 \phantom{= \frac{27}{4}} = & 315 \\ \hline & 321\frac{3}{4} = \text{ Product.} \end{array}$$

Axiom 7.—If any number be both multiplied and divided by the same number, the value of the original number is not changed.

Cor. 1.—If the multiplier is greater than the divisor, the product is greater than the number multiplied, but if the multiplier is less than the divisor, then the product is less than the multiplicand.

Cor. 2.—Multiplying the numerator or dividing the denominator by any number, multiplies the fraction by the same number.

FRACTIONS.

EXAMPLES.

1. Multiply $\frac{1}{2}, \frac{2}{3}, \frac{3}{4}, \frac{4}{5}, \frac{5}{6}, \frac{6}{7}, \frac{7}{8}, \frac{8}{9}, \frac{9}{10}$. Product $= \frac{1}{10}$.
2. Multiply $\frac{1}{2}, \frac{1}{3}, \frac{1}{4}, \frac{1}{5}, \frac{1}{6}$. Product $= \frac{1}{6}$.
3. Multiply $\frac{7}{9}, \frac{9}{11}, \frac{9}{13}$ and $\frac{11}{13}$. Product $= \frac{63}{143}$.
4. Multiply $\frac{1}{3}, \frac{3}{7}, \frac{7}{11}, \frac{11}{13}$ and $\frac{13}{17}$. Product $= \frac{1}{17}$.
5. Multiply $\frac{7}{15} \times \frac{1}{2} \times \frac{4}{7} \times \frac{9}{11} \times \frac{11}{13}$. Product $= \frac{1}{6}$.
6. Multiply $45\frac{1}{2}$ by 15.

$$45\frac{1}{2} \times 15 = \begin{array}{r} 225 \\ 45 \\ \hline 682\frac{1}{2} \end{array} \qquad \begin{array}{c} 7\frac{1}{2} \\ \frac{91}{2} \times 15 = 910 \\ 455 \\ \hline \frac{1365}{2} = 682\frac{1}{2}. \end{array}$$

7. Multiply $74\frac{3}{4}$ by 12.

$$74\frac{3}{4} \times 12 = \frac{299}{4} \times \overset{3}{\cancel{12}} = 897.$$

8. Multiply $27\frac{3}{4}$ by $33\frac{1}{5}$.

$$\frac{111}{4} \times \frac{166}{5} = \frac{18426}{20} = 921\frac{6}{20} = 921\frac{3}{10}.$$

$$\underset{2}{\frac{111}{4}} \times \overset{83}{\underset{5}{\frac{166}{5}}} = \frac{9213}{10} = 921\frac{3}{10} = \text{Product.}$$

9. Multiply $54\frac{3}{4}$ by $37\frac{1}{3} = \frac{219}{4} \times \frac{112}{3} = \frac{24741}{12}$.
10. Multiply $67\frac{2}{3}$ by $51\frac{1}{4}$.
11. Multiply $91\frac{5}{8}$ by $56\frac{3}{4}$.
12. Multiply $5376\frac{7}{8}$ by $821\frac{1}{2}$.
13. Multiply $6274\frac{3}{8}$ by $237\frac{3}{8}$.

Rem.—When one or both factors are mixed numbers, it is generally best to reduce them to improper fractions.

DIVISION OF FRACTIONS.

PROBLEMS.

1. Divide 10 by 5.   *Ans.* $10 \div 5 = \frac{10}{5} = 2$.
2. Divide 6 by 3.   *Ans.* $\frac{6}{3} = 2$.
3. Divide 3 by 3.   *Ans.* $\frac{3}{3} = 1$.
4. Divide 1 by 2.   *Ans.* $\frac{1}{2}$.
5. Divide 1 by 3.   *Ans.* $\frac{1}{3}$.
6. Divide 2 by 3.   *Ans.* $\frac{2}{3}$.
7. Divide 3 by 4.   *Ans.* $\frac{3}{4}$.
8. Divide 5 by 4.   *Ans.* $\frac{5}{4} = 1\frac{1}{4}$.
9. Divide $\frac{1}{2}$ by 2, or divide $\frac{1}{2}$ into two equal parts.
   Ans. $\frac{1}{2 \times 2} = \frac{1}{4}$.
10. Divide $\frac{2}{3}$ by 2, or divide $\frac{2}{3}$ into two equal parts.
   Ans. $\frac{2 \div 2}{3} = \frac{1}{3}$.
11. Divide $\frac{1}{3}$ by 2, or divide $\frac{1}{3}$ into two equal parts.
   Ans. $\frac{1}{3 \times 2} = \frac{1}{6}$.
12. Divide $\frac{2}{3}$ by $\frac{1}{3}$, or how often is $\frac{1}{3}$ contained in $\frac{2}{3}$.
   Ans. Evidently twice.
13. Divide $\frac{1}{2}$ by $\frac{1}{4}$, or how often is $\frac{1}{4}$ contained in $\frac{1}{2} = \frac{2}{4}$.
   Ans. Evidently twice.
14. Divide $\frac{1}{2}$ by $\frac{1}{3}$, or how often is $\frac{1}{3}$ contained in $\frac{1}{2}$.
   Ans. $\frac{1}{2} = \frac{3}{6}$ and $\frac{1}{3} = \frac{2}{6}$; $\frac{3}{6} \div \frac{2}{6} = \frac{3}{2} = 1\frac{1}{2}$.
15. Divide $\frac{1}{2}$ by $\frac{1}{5}$, or how often is $\frac{1}{5}$ contained in $\frac{1}{2}$.
   Ans. $\frac{1}{2} = \frac{5}{10}$ and $\frac{1}{5} = \frac{2}{10}$; $\frac{5}{10} \div \frac{2}{10} = \frac{5}{2}$.

Cor. 1.—The numerator of a fraction is a dividend, the denominator a divisor, and the fraction itself the quotient.

Cor. 2.—To divide a fraction by a fraction, reduce both to a common denominator and then divide the numerator of the dividend by the numerator of the divisor.

Cor. 3.—Dividing the numerator or multiplying the denominator by any number, divides the fraction by the same number.

THEOREM.

To divide any number by a fraction, invert the divisor and multiply by it.

PROBLEMS.

1. Divide 24 by 6.

$$\frac{24}{6} = \frac{2, 2, 2, 3}{2, 3} = 2, 2 = 4.$$

2. Divide 24 by $\frac{2}{3}$.

$$\frac{24}{\frac{2}{3}} = \frac{2, 2, 2, 3}{\frac{2}{3}} = 2, 2, 2 = 8 = \overset{8}{\cancel{24}} \times \frac{3}{\cancel{2}} = 8.$$

In the second division the divisor 6 is itself to be divided by 2; that is, its factor 2 is to be canceled.

Canceling a factor in the divisor multiplies the quotient by the same factor.

3. Divide $\frac{24}{3}$ by $\frac{2}{3}$.

$$\overset{4}{\cancel{\tfrac{24}{3}}} \times \frac{3}{\cancel{2}} = 4, \quad \text{or} \quad \tfrac{24}{3} = \overset{\cdot}{12};$$
and $\quad \frac{2}{3} \doteq 3; \quad \text{and} \quad \tfrac{12}{3} = 4.$

In this case a factor is to be canceled in both dividend and divisor.

Canceling a factor in the dividend divides the quotient by the same factor.

FRACTIONS.

An integral number, or a fraction, is divided by a fraction by inverting the divisor and then multiplying by it.

EXAMPLES.

1. Divide 43 by $\frac{4}{5}$.

$$43 \times \frac{5}{4} = \frac{215}{4} = 53\frac{3}{4}.$$

2. Divide 86 by $\frac{2}{7}$.

$$\overset{43}{\cancel{86}} \times \frac{7}{\cancel{2}} = 301.$$

3. Divide 95 by $\frac{5}{6}$.
4. Divide $115\frac{3}{4}$ by $\frac{4}{5}$.
5. Divide $\frac{4 \cdot 3}{6 \cdot 6}$ by $\frac{8 \cdot 8}{15 \cdot 8}$.

QUESTIONS.

1. One is what part of 2 ? *Ans.* $\frac{1}{2}$.
2. Two is what part of 3 ? *Ans.* $\frac{2}{3}$.
3. Three is what part of 4 ? *Ans.* $\frac{3}{4}$.

4. One-half is what part of $\frac{2}{3}$? *Ans.* $\frac{\frac{1}{2}}{\frac{2}{3}}$.

5. One-third is what part of $\frac{4}{5}$? *Ans.* $\frac{\frac{1}{3}}{\frac{4}{5}}$.

6. Three-fourths is what part of $\frac{7}{8}$? *Ans.* $\frac{\frac{3}{4}}{\frac{7}{8}}$.

COMPLEX FRACTIONS.

When one or both terms of a fraction are either fractions or mixed numbers, it is called a *Complex Fraction;* thus,

$$\frac{3\frac{1}{4}}{5}, \quad \frac{\frac{1}{2}}{\frac{2}{3}}, \quad \frac{3}{\frac{1}{2}}, \quad \frac{37\frac{1}{2}}{62\frac{1}{2}}, \quad \frac{100\frac{1}{2}}{100},$$ etc., are complex fractions.

58 FRACTIONS.

Rem.—When we consider that the numerator of a fraction represents a dividend, and the denominator a divisor, a complex fraction is readily reduced to a simple fraction.

1. Reduce $\dfrac{3\frac{1}{2}}{5}$ to a simple fraction.

$$\dfrac{3\frac{1}{2}}{5} = \dfrac{\frac{7}{2}}{5} = \tfrac{7}{2} \times \tfrac{1}{5} = \tfrac{7}{10}.$$

Rem.—As the denominator is a divisor, it must be inverted.

2. Reduce $\dfrac{\frac{1}{2}}{\frac{2}{3}}$ to a simple fraction.

$$\dfrac{\frac{1}{2}}{\frac{2}{3}} = \tfrac{1}{2} \times \tfrac{3}{2} = \tfrac{3}{4}.$$

3. Reduce $\dfrac{3}{\frac{2}{5}}$ to a simple fraction.

$$\dfrac{3}{\frac{2}{5}} = 3 \times \tfrac{5}{2} = \tfrac{15}{2} = 7\tfrac{1}{2}.$$

4. Reduce $\dfrac{37\frac{1}{2}}{62\frac{1}{2}}$ to a simple fraction.

$$\dfrac{37\frac{1}{2}}{62\frac{1}{2}} = \dfrac{\frac{75}{2}}{\frac{125}{2}} = \tfrac{75}{2} \times \tfrac{2}{125} = \tfrac{3}{5}.$$

5. Multiply $\dfrac{100\frac{1}{4}}{100\frac{1}{2}} \times \dfrac{40\frac{1}{3}}{60\frac{1}{2}} = \dfrac{\frac{201}{2}}{\frac{401}{4}} \times \dfrac{\frac{121}{3}}{\frac{121}{2}} = \tfrac{201}{2} \times \tfrac{121}{3} \times \tfrac{4}{401} \times \tfrac{2}{121} = \tfrac{268}{401}.$

6. Divide $\dfrac{1}{2}$ by $\dfrac{\frac{2}{3}}{\frac{3}{4}} = \dfrac{\frac{1}{2}}{\frac{2}{3}} \times \dfrac{\frac{4}{3}}{1} = \tfrac{1}{2} \times \tfrac{3}{2} \times \tfrac{4}{3} \times \tfrac{1}{1} = \tfrac{1}{1}.$

$\tfrac{1}{2} \times \tfrac{3}{2} \times \tfrac{4}{3} \times \tfrac{1}{1} = \tfrac{1}{1}.$

In changing the division to multiplication, the whole divisor must be inverted; that is, $\frac{3}{4}$ becomes the numerator and $\frac{2}{3}$ becomes the denominator, with the sign of multiplication; then again the two denominators must be inverted; then they are all in the form of multiplication.

FRACTIONS.

PROBLEMS.

1. Reduce $\frac{\frac{1}{2}}{\frac{2}{3}}$ to a simple fraction.

$$\frac{1}{2} \times \frac{3}{2} = \frac{3}{4} \text{ simple fraction.}$$

REM.—It is evident that the above expression indicates that $\frac{1}{2}$ is to be divided by $\frac{2}{3}$.

2. Reduce $\frac{\frac{1}{2}}{\frac{2}{3}} \times \frac{\frac{2}{3}}{\frac{3}{4}}$.

$$\frac{1}{2} \times \frac{3}{2} \times \frac{2}{3} \times \frac{4}{3} = \frac{2}{3}.$$

REM.—$\frac{2}{3}$ and $\frac{3}{4}$ are divisors and must be inverted.

3. Reduce $\frac{\frac{2}{3}}{\frac{3}{4}} \div \frac{\frac{3}{4}}{\frac{4}{5}} = \frac{\frac{2}{3}}{\frac{3}{4}} \times \frac{\frac{4}{5}}{\frac{3}{4}}$.

$$\frac{2}{3} \times \frac{4}{3} \times \frac{4}{3} \times \frac{5}{3} = \frac{2}{15}.$$

REM.—By inverting the whole divisor, problem 3d is changed from division to multiplication.

EXAMPLES.

1. Reduce $\frac{3\frac{1}{2}}{5}$ to a simple fraction.

$$\frac{7}{2} \times \frac{1}{5} = \frac{7}{10} \text{ simple fraction.}$$

2. Reduce $\frac{3}{\frac{2}{5}}$.

$$3 \times \frac{5}{2} = \frac{15}{2} = 7\frac{1}{2}.$$

3. Reduce $\frac{62\frac{1}{2}}{87\frac{1}{2}}$.

$$\tfrac{125}{2} \times \tfrac{2}{175} = \tfrac{5}{7} \text{ simple fraction.}$$

FRACTIONS.

4. Reduce $\dfrac{100\frac{1}{4}}{100\frac{1}{4}}$.

$\dfrac{201}{4} \times \dfrac{4}{401} = \dfrac{404}{404}.$

5. Reduce $\dfrac{3}{4} \times \dfrac{4}{11}$.

6. Reduce $\dfrac{5}{9} \times \dfrac{4}{5}$.

7. Reduce $\dfrac{7}{8} \div \dfrac{8}{11}$.   *Ans.* $\dfrac{11}{8} = 1\frac{1}{18}$.

8. Reduce $\dfrac{21}{13} \div \dfrac{29}{10}$.   *Ans.* 3.

MISCELLANEOUS QUESTIONS.

1. A man spends $\frac{3}{4}$ of his income for board, $\frac{1}{8}$ for clothing, and has 50 dollars left, what is his income?
    Ans. 600 dollars.

2. What number divided by $\frac{2}{3}$ will give for a quotient $\frac{3}{8}$?
    Ans. $\frac{1}{4}$.

3. What number multiplied by $\frac{3}{8}$ will give a product of $\frac{4}{5}$?   *Ans.* $\frac{32}{15}$.

4. In an orchard $\frac{1}{2}$ of the trees bear apples, $\frac{1}{4}$ peaches, $\frac{1}{8}$ plums, and 20 trees bear pears. How many trees in the orchard?   *Ans.* 80 trees.

5. What are the numbers between 15 and 450 which have the former for their greatest common divisor, and the latter for their least common multiple?
    Ans. 30, 45, 75, 90, 150, 225.

6. What six numbers between 35 and 840 have the former for their greatest common divisor, and the latter for their least common multiple?
    Ans. 70, 105, 140, 210, 280, 420.

FRACTIONS.

7. A father bequeathed $\frac{4}{9}$ of his estate to one son, and the rest of his estate to another; the difference of their legacies were 1560 dollars. How much did each receive?
Ans. The one 7800 dollars, and the other 6240 dollars.

8. A cargo is worth five times the value of the ship. What part of the cargo is $\frac{5}{12}$ of the ship and cargo?
Ans. $\frac{1}{2}$.

REM.—2 is what part of 3? *Ans.* $\frac{2}{3}$. 3 is what part of 4? *Ans.* $\frac{3}{4}$. $\frac{1}{2}$ is what part of $\frac{2}{3}$. *Ans.* $\frac{\frac{1}{2}}{\frac{2}{3}} = \frac{1}{2} \times \frac{3}{2} = \frac{3}{4}$.

9. If A can do a piece of work in 4 days, and B can do it in 5 days, what part can each do in a day? What part of it can both do in a day? How many days will it take both to do it?
Ans. A can do $\frac{1}{4}$ in a day, B $\frac{1}{5}$, both $\frac{9}{20}$, and both can do it in $2\frac{2}{9}$ days.

10. A and B bought a barrel of flour, for which A paid 5 dollars, and B paid 6 dollars. What part of the flour should each have? *Ans.* A $\frac{5}{11}$ and B $\frac{6}{11}$.

11. Three men, A, B, and C, bought 640 acres of land. A paid 4000 dollars, B paid 5000 dollars, and C 7000 dollars. How many acres should each receive?
Ans. A 160 acres, B 200 acres, and C 280 acres.

12. A man left $\frac{1}{3}$ of his property to his wife, $\frac{1}{4}$ to each of his two daughters, $\frac{1}{12}$ to his son, and the remainder, which was 500 dollars, to his servant. What was the value of his whole estate? *Ans.* 6000 dollars.

13. If 3650 is $\frac{5}{12}$ of some number, what is $\frac{3}{5}$ of the same number? *Ans.* 5256.

14. A sold $\frac{5}{24}$ of a ship for 3650 dollars, and B owns $\frac{3}{5}$ of the ship. What is it worth at the same rate?
Ans. 5256.

DECIMAL FRACTIONS.

Fractions whose denominators are 10, 100, 1000, etc., are rendered decimals of the same name by a little change in form; thus, a decimal point is placed on the left of the decimals, or on the right of the units, and the same relation exists between the successive orders, as in abstract numbers, but the orders themselves are reversed.

$$\frac{1}{10} = .1, \qquad \frac{1}{1000} = .001,$$
$$\frac{1}{100} = .01, \qquad \frac{1}{10000} = .0001,$$

and are read alike; thus,

one tenth,   one thousandth,
one hundredth,   one ten-thousandth.

Also, $\frac{3}{10} = .3$, read three tenths;

$\frac{7}{100} = .07$, seven hundredths;

$\frac{36}{100} = .36$, thirty-six hundredths;

$\frac{456}{1000} = .456$, four hundred and fifty-six thousandths.

Hence, to enumerate a decimal fraction, read it as you would an integral number, adding to this the name of the denominator, when a common fraction, which will be expressed by 1 with as many zeros attached to it as there are numbers of decimal figures.

ADDITION AND SUBTRACTION.

EXAMPLES.

1. Add .1, .01, .001, .0001, and .0001; thus:

(1.)	(2.)	(3.)
.1	Add .0234	Add 5.634
.01	.213	21.321
.001	.3146	.654
.0001	.32	.012
.00001	.6	5.364
.11111	1.4710	32.985

	(4.)		(5.)
From	4.36215	From	.326159
Take	1.83754	Take	.234573
Rem.	2.52461	Rem.	.091586

Cor.—As the relation of the orders are the same, and the decimals rise in value in the same direction, whilst in name they take the opposite direction; hence, addition and subtraction of decimals are performed as in Integral Numbers.

MULTIPLICATION.

THEOREM.

In the multiplication of decimals, the product will have as many places of decimals as both factors.

$$\frac{1}{10} \times \frac{1}{10} = \frac{1}{100} \quad \therefore \quad .1 \times .1 = .01,$$

and

$$\frac{1}{100} \times \frac{1}{10} = \frac{1}{1000} \quad \therefore \quad .01 \times .1 = .001.$$

64   DECIMAL FRACTIONS.

	1ST COL.		2D COL.
or,	$1 \times 1 = 1$	and	$1 \times .1 = .1$
	$.1 \times 1 = .1$	and	$.1 \times .1 = .01$
	$.01 \times 1 = .01$	and	$.01 \times .1 = .001$

The first column of products is the same as the first column of multiplicands, as 1 is the multiplier. The multiplier in the second case is one-tenth, consequently the products of the second column must be one-tenth of the first.

Therefore the product of two decimal factors will have as many decimal places as both factors.

$$1 \times 1 = 1, \text{ units.}$$
$$.1 \times .1 = .01, \text{ hundredths.}$$
$$.01 \times .01 = .0001, \text{ ten thousandths.}$$

$$1 \times 1 = 1, \text{ units.}$$
$$10 \times 10 = 100, \text{ hundreds.}$$
$$100 \times 100 = 10000, \text{ ten thousands.}$$

REM.—Observe the correspondence in name, when the contrary orders are multiplied.

PROBLEMS.

1. Multiply
```
     3.156
      .215
    ------
    15780
     3156
     6312
    ------
    .678540
```

2. Multiply
```
      .534
      .136
    ------
     3204
     1602
      534
    ------
    .072624
```

REM.—Each product must have six decimals, hence in the second example a zero must be prefixed.

(3.)	(4.)
.01	.00001
.01	.00001
.0001	.000000001

DIVISION.

Corollaries to Theorem, Page 54.

Cor. 1.—As the product of the divisor and quotient is equal to the dividend, therefore the dividend has as many decimal figures as both divisor and quotient.

Cor. 2.—If the divisor has decimal figures and the dividend has none, or less than the divisor, as many must be added to the dividend as to make the number equal to that of the divisor, and then the quotient will be integral. If more decimals are added to the dividend, the quotient will contain as many.

PROBLEMS.

1. Divide 21.4263 by 3.12.

```
3.12 ) 21.42|63 ( 6.86 +
       18 72
       ─────
        2 706
        2 496
        ─────
          2103
          1872
          ────
           231, remainder.
```

As the divisor has two places of decimals, the quotient will be integral for two places of decimals in the dividend; after that the quotient will be decimal.

2. Reduce the fraction ¼ to a decimal.

```
4 ) 1.00
    ────
     .25
```

$$\frac{3}{5}) 3.0$$
$$.6$$

Cor.—Any common fraction may be reduced to a decimal by performing the division indicated by the terms.

EXAMPLES.

1. Multiply 1 by .1; by .01; by .001; by .0001.

2. Multiply .1 by .1; by .01; by .001; by .0001.

3. Multiply .2 by 2; .03 by .4; .05 × .04; .06 × .003; and .003 × .004.

4. Multiply 4.732 by .345.

5. Multiply 2.074 by .021.

6. Multiply 3.541 by .002.

7. Multiply .002 by .3754.

8. Multiply 721.56 by 21.42.

9. Multiply 642.54 by 2162.

10. Multiply 756.48 by 4635.

Rem.—Prove the last seven examples by division.

PRACTICAL EXAMPLES.

1. A merchant sold 205 yards cotton cloth at $.125 per yard, 75 yards gray flannel at $.625 per yard, 12 pairs hose at $.375 per pair, 54 yards linen at $.555 per yard. What was the amount of the bill? *Ans.* $106.97.

2. Bought five tracts of land; viz., 237 acres at $57.43 per acre, 326 acres at $49.02 per acre, 431 acres at $31.21 per acre, 1274 acres at $12.48 per acre, and 21346 acres at $2.045 per acre. The whole is to be paid in three equal instalments; how much is each payment?

Ans. 34198.34\frac{1}{3}$.

This table of aliquot parts enables us to shorten the operations of multiplication and division.

10 cts. = \$$\frac{1}{10}$.	75 cts. = \$$\frac{3}{4}$.	66$\frac{2}{3}$ cts. = \$$\frac{2}{3}$.
20 cts. = \$$\frac{1}{5}$.	12$\frac{1}{2}$ cts. = \$$\frac{1}{8}$.	37$\frac{1}{2}$ cts. = \$$\frac{3}{8}$.
25 cts. = \$$\frac{1}{4}$.	16$\frac{2}{3}$ cts. = \$$\frac{1}{6}$.	62$\frac{1}{2}$ cts. = \$$\frac{5}{8}$.
50 cts. = \$$\frac{1}{2}$.	33$\frac{1}{3}$ cts. = \$$\frac{1}{3}$.	87$\frac{1}{2}$ cts. = \$$\frac{7}{8}$.

DECIMAL FRACTIONS. 67

EXAMPLES.

1. Multiply 576 by 100.   *Ans.* 57600.
2. Multiply 576 by 25.
$$576 \times \frac{100}{4} = 14400.$$
3. Multiply 576 by 50.
$$576 \times \frac{100}{2} = 28800.$$
4. Divide 576 by 100.   *Ans.* 5.76.
5. Divide 576 by 50.
$$576 \div \frac{100}{2} = 576 \times 2 \div 100 = 11.52.$$
6. Divide 576 by 25.
$$576 \div \frac{100}{4} = 576 \times 4 \div 100 = 23.04.$$
7. Divide 67453.2645 by 47.215.

SOLUTION.

```
47.215 ) 67453.2645 ( 1428.2 +
         47215
         ──────
         202182
         188860
         ──────
         133226
          94430
         ──────
         387964
         377720
         ──────
         102445
          94430
         ──────
           8015
```

There must be one decimal in the quotient; for 3 in divisor + 1 in quotient = 4 in dividend.

8. Divide 98637.42598 by 21.798.

9. Divide 7326.4873 by 86.324.

10. Divide 83465.987 by 4365.3315.
Add 1 decimal to dividend.

11. What is the cost of 576 yards at .12½?
$$576 \times \tfrac{1}{8} = \$72.$$

12. What is the cost of 576 yards at .16⅔?
$$576 \times \tfrac{1}{6} = \$96.$$

13. What is the cost of 576 yards at .62½?
$$576 \times \tfrac{5}{8} = 576 \times 5 \div 8 = \$360.$$

14. What is the cost of 576 yards at .87½?
$$576 \times \tfrac{7}{8} = 576 \times 7 \div 8 = \$504.$$

CIRCULATING DECIMALS.

PROBLEMS.

1. In the reduction of common fractions to decimals, when the denominator has no other factor than 2 or 5, or 2 and 5, the decimal will terminate with the number of figures equal to the greatest number of factors of 2 or 5 in the denominator.

2. When the common fraction has any other denominator, the decimal fraction will not terminate; and at some point in the division, the quotient will begin to repeat the same figures; each period of which is called a repetend, and the repetends are called Circulating Decimals.

DECIMAL FRACTIONS.

FORMATION OF REPETENDS.

$\frac{1}{9} = .111$, etc. $= .\dot{1}$; $\frac{2}{9} = .222 = .\dot{2}$;
$\frac{12}{99} = .1212$, etc. $= .\dot{1}\dot{2}$.

REM.—When the repetend has the same figure repeated, a dot is placed over the single figure as above, $.\dot{1}$ and $.\dot{2}$; when the repetend has two or more figures, a dot is placed over the first and last; as $.\dot{1}\dot{2}$.

$\frac{237}{999} = 999\ )\ 237.000\ (\ .237237$, etc. $= .\dot{2}3\dot{7}$
$\phantom{\frac{237}{999} = 999\ )\ }199\ 8$
$\phantom{\frac{237}{999} = 999\ )\ \,}\overline{37\ 20}$
$\phantom{\frac{237}{999} = 999\ )\ \,}29\ 97$
$\phantom{\frac{237}{999} = 999\ )\ \,\,}\overline{7\ 230}$
$\phantom{\frac{237}{999} = 999\ )\ \,\,}6\ 993$
$\phantom{\frac{237}{999} = 999\ )\ \,\,\,}\overline{237}$

Remainder the same as the first dividend; hence, the quotient will repeat.

COR. 1.—A repetend is changed to a common fraction by placing under it, for a denominator, as many nines as there are figures in the repetend.

COR. 2.—When the decimal fraction is partly a common decimal and partly a repetend, it is readily put in the form of a complex fraction, and then may be reduced accordingly; thus,

$.32\dot{5}7\dot{6} = \dfrac{32\frac{576}{999}}{100} = \dfrac{\frac{32 \times 999 + 576}{999}}{100} = \frac{32544}{99900}$.

REM.—Circulating decimals are seldom met with in practice, and the simplest manner to dispose of them is to reduce them to common fractions, and then use them as such.

REVIEW.

NUMERATION, ADDITION AND SUBTRACTION.

1. Express in figures six hundred millions twenty-one thousand one hundred and four.
2. Express in figures the number, three billions seven millions and three.
3. Express in figures thirty-one billions three thousand and twenty-three.
4. Express in figures the number represented by four units of the ninth order, six of the eighth, one of the sixth, four of the third, and seven of the second.
5. On what does the *local* value of a figure depend?
6. Add the following numbers: one billion, thirty-two millions and twenty-five; twenty-one millions eighty-four thousand one hundred and fifteen; three hundred and sixty-five thousand four hundred and seventy-six.
Ans. 1053449616.
7. From forty-three millions six hundred and thirty-one thousand four hundred and twenty-one, take sixteen millions two thousand and two. *Ans.* 27629419.
8. Add one hundred and five thousand three hundred and four; fifty-seven thousand one hundred and two; eighty-six thousand three hundred and ten; thirty-three hundred and nine; one hundred and three.
Ans. 252128.
9. Add two hundred and seven; twenty-one thousand three hundred and one; eight hundred and four

NUMERATION, ADDITION AND SUBTRACTION. 71

thousand and thirty-seven; sixty-one thousand three hundred and four. *Ans.* 886849.

10. A had five thousand dollars, B two thousand five hundred; A lost twelve hundred and seventy-five and B gained thirteen hundred and forty-three dollars. How much had each? *Ans.* A had $3725; B $3843.

11. The minuend is fifty-three thousand two hundred and forty-three; and the difference is seven hundred and eighty-four. What is the subtrahend? *Ans.* 52459.

12. From 43722110523 take 63110421.

13. Name the three terms used in Subtraction. Which is the greatest?

14. What is the minuend? What is the difference between the minuend and the sum of the other two terms?

15. 3 million 225 thousand and 26 plus 12 hundred plus 1 thousand plus 325, minus 1364 = ?
Ans. 3226187.

16. The subtrahend is 572342, and the remainder is 3642, what is the minuend? *Ans.* 575984.

17. A has 5021 acres of land, B has 754 acres more than A, and C has 1012 acres more than both the others. How many have they altogether? *Ans.* 22604 acres.

18. A commenced business with $5000; the first year he gained $2112, the second year $3743, and the third year he lost $5321; how much had he then?
Ans. $5534.

19. A borrowed $7000; in one month he paid $2300, and in another month $1000 more; how much did he still owe? *Ans.* $3700.

20. Two men together have $11364, the one has $3420; how much has the other? *Ans.* $7944.

MULTIPLICATION AND DIVISION.

1. The two factors are 341 and 257; what is the product? *Ans.* 87637.

2. The divisor is 15, the quotient 31, and the remainder 5; what is the dividend? *Ans.* 470.

3. One factor of a number is 12 and the other factor is 2602; what is the number? *Ans.* 31224.

4. The remainder is 14, the quotient 407, and the dividend 61878; what is the divisor? *Ans.* 152.

5. If a ship sail 185 miles in a day, how far will it sail in 51 days? *Ans.* 9435 miles.

6. If the earth moves in its orbit 68000 miles in an hour, how far will it move in 24 hours?
Ans. 1632000 miles.

7. James earned $15 in a week and paid $5 for his board; how much did he save in 52 weeks? *Ans.* $520.

8. A drover bought 45 horses at $84 each, 72 cows at $25 each, 315 sheep at $6 each; what did they all cost?
Ans. $7470.

9. Two men start from the same place and travel in opposite directions; the one goes 5 miles an hour and the other 3; how far apart will they be in 12 hours?
Ans. 96 miles.

10. If the men mentioned in the last example had traveled the same direction for 12 hours, how far apart would they be? *Ans.* 24 miles.

11. The divisor is 8, and the quotient 1142; what is the dividend? *Ans.* 9136.

12. If one sheep cost $5, how many sheep can be bought for $2315? *Ans.* 463 sheep.

13. The salary of the President of the United States is \$50000; what is it per day? *Ans.* \$137.

14. There are 3150 gallons in 50 hogsheads; how many gallons are there in one hogshead?

15. The dividend is 582, the divisor 15, and the remainder 12; what is the quotient? *Ans.* 38.

16. The three factors of a certain number are 35, 43, and 57; what is the number?

17. What is the cost of 554 acres of land, at \$65 an acre?

18. The wine gallon contains 231 cubic inches; how many cubic inches are in a hogshead of 63 gallons?

19. A drover bought 53 horses at \$75 each, 25 cows at \$21 each, 3458 sheep at \$6 each: he sold them for \$251 less than he paid; for how much did he sell them? *Ans.* 24997.

20. What number multiplied by 11 gives 638?

REDUCTION OF FRACTIONS.

Reduce to the least common denominator,

1. $\frac{1}{2}$, $\frac{2}{3}$, $\frac{4}{5}$. *Ans.* $\frac{15}{30}$, $\frac{20}{30}$, $\frac{24}{30}$.
2. $\frac{1}{5}$, $\frac{2}{7}$, $\frac{5}{9}$. *Ans.* $\frac{63}{315}$, $\frac{90}{315}$, $\frac{175}{315}$.
3. $\frac{1}{2}$, $\frac{2}{5}$, $\frac{5}{6}$, $\frac{3}{8}$. *Ans.* $\frac{60}{120}$, $\frac{48}{120}$, $\frac{100}{120}$, $\frac{45}{120}$.
4. $\frac{1}{2}$ of $\frac{2}{3}$, $\frac{3}{4}$ of $\frac{5}{6}$, and $\frac{7}{8}$ of $\frac{2}{9}$. *Ans.* $\frac{24}{72}$, $\frac{45}{72}$, $\frac{14}{72}$.
5. $\frac{1}{2}$ of $2\frac{1}{2}$, $\frac{2}{3}$ of $3\frac{3}{4}$, and $\frac{4}{5}$ of $5\frac{1}{4}$. *Ans.* $\frac{75}{60}$, $\frac{150}{60}$, $\frac{252}{60}$.
6. $\frac{1}{2}$, $\frac{1}{4}$, and $\frac{3}{8}$. *Ans.* $\frac{4}{8}$, $\frac{2}{8}$, $\frac{3}{8}$.
7. $\frac{1}{3}$, $\frac{1}{6}$, and $\frac{5}{12}$. *Ans.* $\frac{4}{12}$, $\frac{2}{12}$, $\frac{5}{12}$.
8. $\frac{1}{4}$, $\frac{3}{5}$, $\frac{7}{10}$, and $\frac{9}{20}$. *Ans.* $\frac{5}{20}$, $\frac{12}{20}$, $\frac{14}{20}$, $\frac{9}{20}$.
9. $\frac{1}{3}$, $\frac{2}{9}$, $\frac{1}{4}$, $\frac{3}{8}$, $\frac{2}{5}$, and $\frac{3}{10}$.
   Ans. $\frac{120}{360}$, $\frac{80}{360}$, $\frac{90}{360}$, $\frac{135}{360}$, $\frac{144}{360}$, $\frac{108}{360}$.
10. $1\frac{1}{4}$, $3\frac{1}{3}$, $4\frac{1}{5}$. *Ans.* $\frac{75}{60}$, $\frac{200}{60}$, $\frac{252}{60}$.

FRACTIONS.

ORAL EXERCISES.

1. What is $\frac{1}{2}$ of $\frac{2}{3}$? $\frac{1}{2}$ of $\frac{3}{4}$? $\frac{1}{3}$ of $\frac{1}{4}$? $\frac{1}{4}$ of $\frac{4}{5}$? $\frac{1}{5}$ of $\frac{5}{6}$? $\frac{1}{7}$ of $\frac{7}{8}$?

2. One-third of 6 is $\frac{1}{2}$ of what number? $\frac{1}{3}$? $\frac{1}{4}$? $\frac{1}{5}$? $\frac{1}{6}$? $\frac{1}{7}$? $\frac{1}{8}$? $\frac{1}{9}$? $\frac{1}{10}$? $\frac{1}{11}$? $\frac{1}{12}$?

3. One-third of 9 is $\frac{1}{2}$ of what number? $\frac{1}{3}$? $\frac{1}{4}$? $\frac{1}{5}$? $\frac{1}{6}$? $\frac{1}{7}$? $\frac{1}{8}$? $\frac{1}{9}$? $\frac{1}{10}$? $\frac{1}{11}$? $\frac{1}{12}$?

4. Two-thirds of 9 is $\frac{1}{2}$ of what number? $\frac{1}{3}$? $\frac{1}{4}$? $\frac{1}{5}$? $\frac{1}{6}$? $\frac{1}{7}$? $\frac{1}{8}$? $\frac{1}{9}$? $\frac{1}{10}$? $\frac{1}{11}$? $\frac{1}{12}$?

5. One-third of 12 is $\frac{1}{2}$ of what number? $\frac{1}{3}$? $\frac{1}{4}$? $\frac{1}{5}$? $\frac{1}{6}$? $\frac{1}{7}$? $\frac{1}{8}$? $\frac{1}{9}$? $\frac{1}{10}$? $\frac{1}{11}$? $\frac{1}{12}$?

6. Two-thirds of 12 is $\frac{1}{2}$ of what number? $\frac{1}{3}$? $\frac{1}{4}$? $\frac{1}{5}$? $\frac{1}{6}$? $\frac{1}{7}$? $\frac{1}{8}$? $\frac{1}{9}$? $\frac{1}{10}$? $\frac{1}{11}$? $\frac{1}{12}$?

7. One-fourth of 12 is $\frac{1}{2}$ of what number? $\frac{1}{3}$? $\frac{1}{4}$? $\frac{1}{5}$? $\frac{1}{6}$? $\frac{1}{7}$? $\frac{1}{8}$? $\frac{1}{9}$? $\frac{1}{10}$? $\frac{1}{11}$? $\frac{1}{12}$?

8. One-half of 12 is $\frac{1}{3}$ of what number? $\frac{1}{4}$? $\frac{1}{5}$? $\frac{1}{6}$? $\frac{1}{7}$? $\frac{1}{8}$? $\frac{1}{9}$? $\frac{1}{10}$? $\frac{1}{11}$? $\frac{1}{12}$?

9. Three-fourths of 12 is $\frac{1}{2}$ of what number? $\frac{1}{3}$? $\frac{1}{4}$? $\frac{1}{5}$? $\frac{1}{6}$? $\frac{1}{7}$? $\frac{1}{8}$? $\frac{1}{9}$? $\frac{1}{10}$? $\frac{1}{11}$? $\frac{1}{12}$?

10. One-fourth of 16 is $\frac{1}{2}$ of what number? $\frac{1}{3}$? $\frac{1}{5}$? $\frac{1}{6}$? $\frac{1}{7}$? $\frac{1}{8}$? $\frac{1}{9}$? $\frac{1}{10}$? $\frac{1}{11}$? $\frac{1}{12}$?

11. One-half of 16 is $\frac{1}{3}$ of what number? $\frac{1}{4}$? $\frac{1}{5}$? $\frac{1}{6}$? $\frac{1}{7}$? $\frac{1}{8}$? $\frac{1}{9}$? $\frac{1}{10}$? $\frac{1}{11}$? $\frac{1}{12}$?

12. Three-fourths of 16 is $\frac{1}{2}$ of what number? $\frac{1}{3}$? $\frac{1}{4}$? $\frac{1}{5}$? $\frac{1}{6}$? $\frac{1}{7}$? $\frac{1}{8}$? $\frac{1}{9}$? $\frac{1}{10}$? $\frac{1}{11}$? $\frac{1}{12}$?

13. One-third of 18 is $\frac{1}{2}$ of what number? $\frac{1}{4}$? $\frac{1}{5}$? $\frac{1}{6}$? $\frac{1}{7}$? $\frac{1}{8}$? $\frac{1}{9}$? $\frac{1}{10}$? $\frac{1}{11}$? $\frac{1}{12}$?

14. Two-thirds of 18 is $\frac{1}{2}$ of what number? $\frac{1}{3}$? $\frac{1}{4}$? $\frac{1}{5}$? $\frac{1}{6}$? $\frac{1}{7}$? $\frac{1}{8}$? $\frac{1}{9}$? $\frac{1}{10}$? $\frac{1}{11}$? $\frac{1}{12}$?

15. What part of 1 is $\frac{1}{3}$? $\frac{2}{3}$? $\frac{3}{4}$? $\frac{4}{5}$?

16. What part of 2 is $\frac{1}{3}$? $\frac{2}{3}$? $\frac{3}{4}$? $\frac{4}{5}$?

Ans. $\frac{1}{6}$, $\frac{2}{6}$, $\frac{3}{8}$, $\frac{4}{10}$.

17. What part of 3 is $\frac{1}{2}$? $\frac{1}{3}$? $\frac{2}{3}$? $\frac{3}{4}$? $\frac{4}{5}$?

Ans. $\frac{1}{6}$, $\frac{1}{9}$, $\frac{2}{9}$, $\frac{3}{12}$, $\frac{4}{15}$.

18. What part of $\frac{1}{2}$ is $\frac{1}{3}$? $\frac{2}{3}$? $\frac{3}{4}$? $\frac{4}{5}$?

Ans. $\frac{2}{3}$, $\frac{4}{3}$, $\frac{6}{4}$, $\frac{8}{5}$.

19. What part of $\frac{5}{6}$ is $\frac{1}{2}$? $\frac{1}{3}$? $\frac{2}{3}$? $\frac{3}{4}$? $\frac{4}{5}$?

Ans. $\frac{3}{5}$, $\frac{2}{5}$, $\frac{4}{5}$, $\frac{9}{10}$, $\frac{24}{25}$.

20. What part of $\frac{6}{7}$ is $\frac{1}{2}$? $\frac{1}{3}$? $\frac{2}{3}$? $\frac{3}{4}$? $\frac{4}{5}$?

Ans. $\frac{7}{12}$, $\frac{7}{18}$, $\frac{7}{9}$, $\frac{21}{24}$, $\frac{14}{15}$.

21. What part of 5 is $\frac{1}{2}$? $\frac{1}{3}$? $\frac{2}{3}$? $\frac{3}{4}$? $\frac{4}{5}$?

Ans. $\frac{1}{10}$, $\frac{1}{15}$, $\frac{2}{15}$, $\frac{3}{20}$, $\frac{4}{25}$.

22. What part of 20 is 4? 5? $6\frac{2}{3}$? 10? 15?

Ans. $\frac{1}{5}$, $\frac{1}{4}$, $\frac{1}{3}$, $\frac{1}{2}$, $\frac{3}{4}$.

23. What part of 25 is 4? 5? $6\frac{1}{4}$? 10? $12\frac{1}{2}$?

Ans. $\frac{4}{25}$, $\frac{1}{5}$, $\frac{1}{4}$, $\frac{2}{5}$, $\frac{1}{2}$.

24. What part of 100 is 4? 5? 6? 10? $12\frac{1}{2}$?

Ans. $\frac{1}{25}$, $\frac{1}{20}$, $\frac{3}{50}$, $\frac{1}{10}$, $\frac{1}{8}$.

25. What part of $5\frac{1}{2}$ is $\frac{1}{2}$? $1\frac{1}{2}$? $2\frac{1}{2}$? $3\frac{1}{2}$? $4\frac{1}{2}$?

Ans. $\frac{1}{11}$, $\frac{3}{11}$, $\frac{5}{11}$, $\frac{7}{11}$, $\frac{9}{11}$.

26. What part of $6\frac{1}{4}$ is $\frac{1}{4}$? $\frac{3}{4}$? $1\frac{1}{4}$? $1\frac{1}{2}$? $1\frac{3}{4}$?

Ans. $\frac{1}{25}$, $\frac{3}{25}$, $\frac{5}{25}$, $\frac{6}{25}$, $\frac{7}{25}$.

27. 1 is $\frac{1}{7}$ of what number? $\frac{2}{7}$? $\frac{3}{7}$? $\frac{4}{7}$? $\frac{5}{7}$?

Ans. 7, $3\frac{1}{2}$, $2\frac{1}{3}$, $1\frac{3}{4}$, $1\frac{2}{5}$.

28. 5 is $\frac{1}{7}$ of what number? $\frac{2}{7}$? $\frac{3}{7}$? $\frac{4}{7}$? $\frac{5}{7}$?

Ans. 35, $17\frac{1}{2}$, $11\frac{2}{3}$, $8\frac{3}{4}$, 7.

29. 6 is $\frac{1}{7}$ of what number? $\frac{2}{7}$? $\frac{3}{7}$? $\frac{4}{7}$? $\frac{5}{7}$?

Ans. 42, 21, 14, $10\frac{1}{2}$, $8\frac{2}{5}$.

30. 7 is $\frac{1}{7}$ of what number? $\frac{2}{7}$? $\frac{3}{7}$? $\frac{4}{7}$? $\frac{5}{7}$?

Ans. 49, $24\frac{1}{2}$, $16\frac{1}{3}$, $12\frac{1}{4}$, $9\frac{4}{5}$.

31. $\frac{1}{7}$ of 14 is $\frac{1}{2}$ of what number? $\frac{1}{3}$? $\frac{1}{4}$? $\frac{1}{5}$? $\frac{1}{6}$?

Ans. 4, 6, 8, 10, 12.

32. $\frac{2}{7}$ of 14 is $\frac{1}{2}$ of what number? $\frac{1}{3}$? $\frac{1}{4}$? $\frac{1}{5}$? $\frac{1}{6}$?

Ans. 8, 12, 16, 20, 24.

33. $\frac{3}{7}$ of 14 is $\frac{1}{2}$ of what number? $\frac{1}{3}$? $\frac{1}{4}$? $\frac{1}{5}$? $\frac{1}{6}$?

Ans. 12, 18, 24, 30, 36.

34. $\frac{3}{5}$ of 15 is $\frac{1}{2}$ of what number? $\frac{1}{3}$? $\frac{1}{4}$? $\frac{1}{5}$? $\frac{1}{6}$?

Ans. 18, 27, 36, 45, 54.

35. $\frac{4}{5}$ of 15 is $\frac{1}{2}$ of what number? $\frac{1}{3}$? $\frac{1}{4}$? $\frac{1}{5}$? $\frac{1}{6}$?

Ans. 24, 36, 48, 60, 72.

MULTIPLICATION OF FRACTIONS.

PROBLEMS.

1. Multiply $\frac{1}{2}$ by 1.

$\frac{1}{2} \times 1 = \frac{1}{2}$; and by alternating, $1 \times \frac{1}{2} = \frac{1}{2}$.

2. Multiply $\frac{1}{3}$ by 1.

$\frac{1}{3} \times 1 = \frac{1}{3}$; and by alternating, $1 \times \frac{1}{3} = \frac{1}{3}$.

3. Multiply $\frac{1}{4}$ by 1.

$\frac{1}{4} \times 1 = \frac{1}{4}$; and by alternating, $1 \times \frac{1}{4} = \frac{1}{4}$.

4. Multiply $\frac{1}{2}$ by 2.

$\frac{1}{2} \times 2 = \frac{2}{2}$; and by alternating, $2 \times \frac{1}{2} = \frac{2}{2}$.

5. Multiply $\frac{1}{3}$ by 2.

$\frac{1}{3} \times 2 = \frac{2}{3}$; and by alternating, $2 \times \frac{1}{3} = \frac{2}{3}$.

6. Multiply $\frac{2}{3}$ by 2.

$\frac{2}{3} \times 2 = \frac{4}{3}$; and by alternating, $2 \times \frac{2}{3} = \frac{4}{3}$.

7. Multiply $\frac{2}{5}$ by 2.

$\frac{2}{5} \times 2 = \frac{4}{5}$; and by alternating, $2 \times \frac{2}{5} = \frac{4}{5}$.

DIVISION OF FRACTIONS.

8. Multiply $\frac{1}{4}$ by $\frac{1}{2}$. $\frac{1}{4} \times \frac{1}{2} = \frac{1}{8}$.
9. Multiply $\frac{1}{3}$ by $\frac{1}{3}$. $\frac{1}{3} \times \frac{1}{3} = \frac{1}{9}$.
10. Multiply $\frac{2}{3}$ by $\frac{1}{2}$. $\frac{2}{3} \times \frac{1}{2} = \frac{2}{6}$.
11. Multiply $\frac{2}{3}$ by $\frac{3}{4}$. $\frac{2}{3} \times \frac{3}{4} = \frac{6}{12}$.

REM.—Multiplying 1 by $\frac{1}{2}$ is the same as dividing it by 2, and multiplying it by $\frac{1}{3}$ is the same as dividing it by 3; hence, to multiply a unit by a fraction whose numerator is unity, the product will have unity for its numerator, and the denominator of the fraction for its denominator; if both factors are fractional, the product of their denominators is the denominator of the product; if the numerator of either factor is increased, the product is similarly increased.

COR.—The product of two or more fractions will have for its numerator the product of all the numerators, and for its denominator the product of all the denominators.

DIVISION OF FRACTIONS.

PROBLEMS.

1. Divide $\frac{4}{5}$ by $\frac{1}{5}$.   *Ans.* 4.
2. Divide $\frac{7}{8}$ by $\frac{1}{8}$.   *Ans.* 7.
3. Divide $\frac{5}{6}$ by $\frac{1}{6}$.   *Ans.* 5.
4. Divide $\frac{1}{2}$ by $\frac{1}{4}$.   *Ans.* 2.
5. Divide $\frac{1}{4}$ by $\frac{1}{8}$.   *Ans.* 2.
6. Divide $\frac{1}{4}$ by $\frac{1}{2}$.   *Ans.* $\frac{1}{2}$.
7. Divide 1 by $\frac{1}{2}$; $\frac{1}{3}$; $\frac{1}{4}$; $\frac{1}{5}$.   *Ans.* 2, 3, 4, 5.
8. Divide 1 by $\frac{2}{3}$; $\frac{3}{4}$; $\frac{4}{5}$.   *Ans.* $\frac{3}{2}$, $\frac{4}{3}$, $\frac{5}{4}$.
9. Divide 2 by $\frac{1}{2}$; $\frac{1}{3}$; $\frac{1}{4}$; $\frac{1}{5}$.   *Ans.* 4, 6, 8, 10.
10. Divide 2 by $\frac{2}{3}$; $\frac{3}{4}$; $\frac{4}{5}$.   *Ans.* $\frac{6}{2}$, $\frac{8}{3}$, $\frac{10}{4}$.
11. Divide $\frac{1}{2}$ by $\frac{1}{2}$; $\frac{1}{3}$; $\frac{1}{4}$; $\frac{1}{5}$.   *Ans.* $\frac{2}{2}$, $\frac{3}{2}$, $\frac{4}{2}$, $\frac{5}{2}$.
12. Divide $\frac{1}{2}$ by $\frac{2}{3}$; $\frac{3}{4}$; $\frac{4}{5}$.   *Ans.* $\frac{3}{4}$, $\frac{4}{6}$, $\frac{5}{8}$.

DIVISION OF FRACTIONS.

PRIN. 1.—When the dividend and divisor have a common denominator, divide the numerator of the dividend by the numerator of the divisor.

PRIN. 2.—Any fraction having unity for its numerator, is contained in unity as many times as there are units in its denominator; if the fraction has 2 for its numerator, it is contained one-half as many times; if it has 3 for its numerator, one-third as many times; etc.

PRIN. 3.—Two fractions, each having unity for its numerator, are divided by making the denominator of the divisor the dividend, and that of the dividend the divisor, regarding these numbers as integers and performing the division.

For, $\frac{1}{2} \div \frac{1}{4} = 2$; $\frac{1}{2} \div \frac{1}{6} = 3$; $\frac{1}{2} \div \frac{1}{8} = 4$; $\frac{1}{2} \div \frac{1}{10} = 5$.

$\frac{1}{3} \div \frac{1}{12} = 4$; $\frac{1}{3} \div \frac{1}{18} = 6$; $\frac{1}{3} \div \frac{1}{24} = 8$; $\frac{1}{3} \div \frac{1}{30} = 10$.

$\frac{1}{4} \div \frac{1}{4} = 1$; $\frac{1}{4} \div \frac{1}{6} = \frac{6}{4} = 1\frac{1}{2}$; $\frac{1}{4} \div \frac{1}{8} = \frac{8}{4} = 2$;

$\frac{1}{4} \div \frac{1}{9} = \frac{9}{4} = 2\frac{1}{4}$.

PRIN. 4.—If the dividend is increased, the quotient is increased just as many times; if the dividend is diminished, the quotient is diminished in like manner.

COR.—To divide any number, either integral or fractional, by a fraction, invert the divisor and then multiply the dividend by it.

ADDITION AND SUBTRACTION OF FRACTIONS.

Add
1. $\frac{1}{2}, \frac{2}{3}, \frac{4}{5}, \frac{4}{5}$.   *Ans.* $2\frac{4}{5}\frac{3}{0}$.
2. $\frac{1}{5}, \frac{4}{5}, \frac{7}{15}$, and $\frac{4}{11}$.   *Ans.* $1\frac{7}{8}\frac{7}{5}$.
3. $\frac{1}{5}, \frac{1}{4}, \frac{7}{13}$.   *Ans.* $\frac{3}{4}\frac{3}{0}$.
4. $\frac{1}{3}, \frac{3}{5}, \frac{4}{5}$, and $\frac{5}{6}$.   *Ans.* $2\frac{1}{3}\frac{3}{0}$.
5. $1\frac{1}{3}$, and $3\frac{1}{5}$.   *Ans.* $4\frac{8}{15}$.
6. $2\frac{1}{4}, 3\frac{1}{3}$, and $5\frac{3}{5}$.   *Ans.* $10\frac{1}{2}\frac{3}{0}$.
7. $1\frac{1}{2}, 2\frac{1}{3}, 3\frac{1}{4}$, and $4\frac{1}{5}$.   *Ans.* $11\frac{1}{6}\frac{7}{0}$.
8. $\frac{1}{2}$ of $\frac{2}{3}, \frac{1}{3}$ of $\frac{3}{4}$, and $\frac{1}{4}$ of $\frac{4}{5}$.   *Ans.* $\frac{47}{60}$.
9. $\frac{3}{5}, \frac{4}{5}$, and $\frac{7}{15}$.   *Ans.* $1\frac{5}{7}\frac{2}{5}$.
10. $2\frac{1}{4}, 4\frac{3}{10}$, and $7\frac{7}{80}$.   *Ans.* $13\frac{1}{8}\frac{1}{0}$.
11. From $\frac{3}{4}$ take $\frac{5}{8}$.   *Ans.* $\frac{1}{8}$.
12. From $\frac{4}{5}$ take $\frac{3}{15}$.   *Ans.* $\frac{3}{5}$.
13. From $\frac{5}{11}$ take $\frac{1}{4}$.   *Ans.* $\frac{9}{44}$.
14. From $\frac{7}{15}$ take $\frac{3}{11}$.   *Ans.* $\frac{32}{135}$.
15. From $\frac{5}{11}$ take $\frac{5}{12}$.   *Ans.* $\frac{5}{132}$.
16. What is $12\frac{1}{2} - 7\frac{1}{4}$?   *Ans.* $5\frac{1}{4}$.
17. What is $10\frac{1}{2} - 4\frac{3}{4}$?   *Ans.* $5\frac{3}{4}$.
18. What is $21\frac{2}{3} - 10\frac{1}{4}$?   *Ans.* $10\frac{5}{6}$.
19. What is $25\frac{4}{11} - 12\frac{7}{11}$?   *Ans.* $12\frac{9}{11}$.
20. What is $36\frac{1}{2} - 21\frac{3}{7}$?   *Ans.* $15\frac{1}{11}$.
21. What is $75\frac{4}{25} - 26\frac{1}{4}$?   *Ans.* $48\frac{44}{50}$.
22. What is $\frac{1}{2}$ of $\frac{3}{11} - \frac{1}{3}$ of $\frac{2}{3}$?   *Ans.* $\frac{89}{195}$.
23. What is $\frac{1}{3}$ of $\frac{3}{7} - \frac{1}{4}$ of $\frac{3}{12}$?   *Ans.* $\frac{27}{840}$.
24. What is $\frac{1}{4}$ of $5\frac{3}{4} - \frac{4}{5}$ of $3\frac{1}{2}$?   *Ans.* $\frac{1}{7}$.
25. What is $\frac{4}{5}$ of $\frac{4}{5} - \frac{1}{3}$ of $\frac{1}{3}$?   *Ans.* $\frac{11}{17}$.
26. What is the difference between $\frac{1}{2}$ of $\frac{3}{4}$ of $\frac{4}{5}$ and $\frac{1}{3}$ of $\frac{4}{5}$ of $\frac{1}{2}$?   *Ans.* $\frac{71}{801}$.

27. What is the difference between $\frac{1}{4}$ of $2\frac{2}{3}$ and $\frac{1}{4}$ of $3\frac{1}{4}$? Ans. $\frac{1}{12}$.

What is the difference between
28. $\frac{1}{4}$ of $12\frac{3}{4}$ and $\frac{1}{4}$ of $16\frac{1}{4}$? Ans. $\frac{1}{16}$.
29. $\frac{1}{4}$ of $20\frac{2}{4}$ and $\frac{1}{4}$ of $36\frac{8}{11}$? Ans. $1\frac{17}{44}$.
30. What is the difference between two fractions whose numerators are each 1 and whose denominators are 5 and 7 ? Ans. $\frac{2}{35}$.

Rem.—The numerator will be the difference and the denominator the product of the denominators.

MULTIPLICATION AND DIVISION OF FRACTIONS.

1. Multiply $\frac{2}{9}$ by 3. Ans. $\frac{2}{3}$.
2. Multiply $\frac{1}{4}$ by 4. Ans. 1.
3. Multiply $\frac{2}{15}$ by 5. Ans. $\frac{2}{3}$.
4. Multiply $\frac{4}{18}$ by 6. Ans. $\frac{4}{3}$.
5. Multiply $\frac{1}{2}$ of 8 by $\frac{1}{4}$ of 10. Ans. 4.
6. Multiply $\frac{1}{2}$ of $6\frac{1}{4}$ by $\frac{2}{3}$ of $5\frac{1}{4}$. Ans. $6\frac{41}{100}$.
7. Multiply $\frac{2}{3}$ of $\frac{3}{4}$ of $5\frac{2}{3}$ by $\frac{2}{3}$ of $4\frac{1}{4}$. Ans. $6\frac{23}{48}$.
8. Multiply $6\frac{1}{2}, 5\frac{1}{4}, 3\frac{2}{3}$ and $\frac{2}{3}$ of $5\frac{2}{3}$. Ans. $269\frac{3}{110}$.
9. Multiply $\frac{2}{3}, \frac{3}{4}, \frac{2}{11}, \frac{1}{2}$ of $2\frac{1}{4}$, and $\frac{2}{3}$ of $5\frac{1}{4}$. Ans. $\frac{27}{140}$.
10. Multiply $\frac{2}{3}, \frac{4}{5}, \frac{1}{4}, \frac{7}{8}$, and $\frac{2}{11}$. Ans. $\frac{21}{44}$.
11. Multiply $3\frac{1}{2}, 5\frac{1}{2}, 6\frac{2}{3}, \frac{1}{2}$ of $\frac{4}{11}$, and $7\frac{1}{4}$. Ans. $35\frac{2}{3}$.
12. Divide 7 by $\frac{2}{7}$. Ans. $24\frac{1}{2}$.
13. Divide 9 by $\frac{3}{8}$. Ans. 24.
14. Divide 30 by $2\frac{1}{2}$. Ans. 12.
15. Divide $\frac{1}{2}$ by $\frac{1}{3}$. Ans. $1\frac{1}{2}$.
16. Divide $\frac{9}{10}$ by $\frac{3}{5}$. Ans. $1\frac{1}{2}$.
17. Divide $\frac{3}{8}$ by $\frac{3}{16}$. Ans. 2.
18. Divide $\frac{3}{8}$ by $\frac{2}{15}$. Ans. 3.

19. Divide $\frac{15}{56}$ by $\frac{5}{112}$. *Ans.* 6.
20. Divide $\frac{7}{144}$ by $\frac{9}{1728}$. *Ans.* $9\frac{1}{3}$.
21. Divide $\frac{17}{25}$ by $\frac{37}{45}$. *Ans.* $\frac{153}{185}$.
22. Divide $\frac{547}{650}$ by $\frac{321}{1300}$. *Ans.* $3\frac{131}{321}$.
23. Divide $\frac{1}{3}$ of $\frac{2}{5}$ of $\frac{3}{8}$ of $\frac{4}{9}$ by $\frac{1}{6}$ of $\frac{4}{11}$ of $\frac{9}{10}$. *Ans.* $\frac{11}{27}$.

Reduce to a simple fraction,

24. $\dfrac{\frac{2}{3}}{\frac{3}{4}}$. *Ans.* $\frac{8}{9}$.

25. $\dfrac{\frac{1}{4}}{\frac{3}{8}}$. *Ans.* $\frac{2}{3}$.

26. $\dfrac{\frac{4}{15}}{\frac{7}{25}}$. *Ans.* $\frac{20}{21}$.

27. $\dfrac{\frac{7}{8}\times\frac{9}{14}}{\frac{12}{15}\times\frac{5}{6}}$. *Ans.* $\frac{27}{32}$.

28. $\dfrac{\frac{5}{12}\times\frac{6}{5}}{\frac{2}{5}\times\frac{1}{8}}$. *Ans.* 10.

29. $\dfrac{\frac{5}{6}}{\frac{10}{27}}\div\dfrac{\frac{18}{25}}{\frac{12}{35}}$. *Ans.* $\frac{5}{7}$.

30. $\dfrac{\frac{1}{4}+\frac{2}{3}}{\frac{1}{2}-\frac{1}{4}}\div\dfrac{\frac{7}{8}-\frac{4}{5}}{\frac{7}{12}+\frac{1}{3}}$. *Ans.* $\frac{1210}{27}$.

31. $\dfrac{\frac{2}{3}\times\frac{3}{4}}{\frac{7}{8}\div\frac{1}{3}}\div\dfrac{\frac{3}{8}\div\frac{3}{4}}{\frac{1}{7}\times\frac{2}{3}}$. *Ans.* $\frac{8}{147}$.

32. $\frac{1}{2}\times\frac{2}{3}\times\frac{3}{4}\times\frac{4}{5}\times\frac{5}{6}\div\frac{1}{4}\times\frac{5}{6}\times\frac{3}{8}\times\frac{2}{5}\times\frac{1}{6}$. *Ans.* 32.
33. $37\frac{3}{4}\div 25\frac{5}{8}$. *Ans.* $\frac{302}{205}$.
34. $201\frac{3}{4}\div 307\frac{1}{5}$. *Ans.* $\frac{1345}{2048}$.
35. $\frac{1}{1000}\div 1000$. *Ans.* $\frac{1}{1000000}$.

NUMERATION, ADDITION AND SUBTRACTION OF DECIMALS.

Write in figures,

1. Twelve and five tenths.
2. Five hundredths.
3. Five thousandths.
4. Five hundred, and five hundredths.

5. Five thousand, and five thousandths.
6. Fifty thousand, and five ten-thousandths.
7. Three hundred and seventy-five thousandths.
8. Two thousand three hundred and seventy-five ten-thousandths.
9. Thirty-two thousand three hundred and seventy-five hundred-thousandths.
10. Two hundred and seventy-five hundred-thousandths.
11. Seventy-five hundred-thousandths.
12. Five hundred thousandths.

Read the following :
13. .01, 2.07, 11.11, 13.0013, 50.501, .0016.
14. 33.00033, 119.019, 3.000003, 4.00044.

Add the following numbers:
1. Fifty-five thousand, and twelve thousandths; twenty-one hundred, and twenty-one hundredths; fifty-two, and two hundred and seventy-five thousandths.
Ans. 57152.497.

2. Twenty-one, and twenty-one hundredths; two hundred and ten, and two hundred and ten thousandths ; three thousand four hundred and fifty-one, and three thousand four hundred and fifty-one ten-thousandths.
Ans. 3682.7651.

3. Forty-four, and four tenths; four hundred, and four hundredths; four thousand, and four thousandths; forty thousand, and four ten-thousandths; four hundred thousand, and four hundred thousandths; four million, and four millionths. *Ans.* 4444444.444444.

4. From four million, and four millionths take four thousand, and four thousandths. *Ans.* 3995999.996004.

MULTIPLICATION AND DIVISION OF DECIMALS.   83

  5. From three thousand, and three thousandths take three hundred, and three hundredths. *Ans.* 2699.973.
  6. From 3471.3471 take 325.325.   *Ans.* 3146.0221.
  7. From 3004.3004 take 305.305.   *Ans.* 2698.9954.
  8. From 3000004.3000004 take 20001.20001.
                                Ans. 2980003.0999904.
  9. From 4001.4001 take 401.401.   *Ans.* 3599.9991.

MULTIPLICATION AND DIVISION OF DECIMALS.

1. Multiply 401.401 by 30.30.   *Ans.* 12162.45030.
2. Multiply 4001.4001 by 401.401.
                             Ans. 1606166.0015401.
3. Multiply 2000001.2000001 by 2001.2001.
                    Ans. 4002402601.44032012001.
4. Divide .8 by .4.   *Ans.* 2.
5. Divide .08 by .04.   *Ans.* 2.
6. Divide .008 by .004.   *Ans.* 2.
7. Divide .576 by .024; by .24.   *Ans.* 24 and 2.4.
8. Divide .036 by .12; by .006; by .0006.
                             Ans. .3; 6; 60.
9. Divide .225 by .9; by .003; by .025.
                            Ans. .25; 75; 9.
10. Divide 1.11 by .1; by .001; by .00001.
                        Ans. 11.1; 1110; 111000.
11. Divide 3.43 by .7; by .07; by .0014.
                        Ans. 4.9 : 49; 2450.
12. Divide 271.26 by .003; by .9; by .0009.
                      Ans. 90420; 301.4; 301400.
13. Divide .000121 by 1.1; by .011; by .00011.
                       Ans. .00011; .011; 1.1.

FRACTIONS.

PRACTICAL EXAMPLES.

1. The dividend is $\frac{9}{11}$ and the quotient $\frac{4}{7}$; what is the divisor? *Ans.* $1\frac{8}{15}$.

2. The product of two numbers is $3\frac{4}{7}$ and one of the numbers is $5\frac{1}{3}$; what is the other number? *Ans.* $\frac{15}{22}$.

3. If 2 be added to each term of $\frac{1}{5}$, will the fraction be increased or diminished? *Ans.* Increased by $\frac{3}{10}$.

4. If 2 be added to each term of $\frac{6}{5}$, will the fraction be increased or diminished? *Ans.* Diminished by $\frac{2}{35}$.

Cor.—If both terms of a proper fraction be increased by the same number, the value of the fraction is increased; but if both terms of an improper fraction be similarly increased, the value is diminished.

5. Received from one man $56\frac{2}{3}$, from a second one $95\frac{1}{4}$, and then paid a third $124\frac{2}{15}$; how much remained? *Ans.* $28\frac{1}{4}$.

6. A gentleman bought a horse, buggy, and harness for $360; the harness cost $\frac{1}{3}$ as much as the buggy, and the buggy one-half as much as the horse; what did each cost? *Ans.* Harness $36, buggy $108, horse $216.

7. A and B together have $204; and $\frac{2}{3}$ of A's is equal to $\frac{3}{4}$ of B's; how much has each? *Ans.* A has $108, B $96.

FRACTIONS.

$\frac{2}{3}$ A's = $\frac{3}{4}$ B's;
$\frac{8}{12}$ A's = $\frac{9}{12}$ B's;
8 times A's = 9 times B's.

Therefore, A has 9 as often as B has 8; and the two have 17 twelve times in 204; hence,

A has 9 × 12 = 108,
and   B has 8 × 12 = 96.

8. A, B, and C bought a farm for $6000. A paid $\frac{2}{5}$, B $\frac{1}{3}$, and C the balance; what did each pay?
Ans. A $2400, B $2000, C $1600.

9. If 3$\frac{1}{2}$ tons of coal cost 24\frac{1}{2}$, what will 25$\frac{2}{5}$ tons cost?   *Ans.* $178.

10. If 5$\frac{2}{3}$ cords of wood cost 25\frac{1}{2}$, what will 1 cord cost? What will 26$\frac{2}{3}$ cords cost? What will 57$\frac{1}{4}$ cords cost?   *Ans.* 4\frac{1}{2}$, $120, and 257\frac{2}{3}$.

11. If 9$\frac{1}{4}$ barrels of flour cost 51\frac{1}{4}$, what cost 1 barrel? What cost 38$\frac{2}{3}$ barrels?   *Ans.* 5\frac{1}{4}$ and 212\frac{2}{3}$.

12. A can dig a ditch in 6 days, B in 9 days; how much will each do in 1 day? How much will both do in 1 day? How long will it take the two to do the work?

Ans. A can do $\frac{1}{6}$ and B $\frac{1}{9}$ in a day = both $\frac{5}{18}$ in a day, and both can do it in 3$\frac{3}{5}$ days.

13. Divide $60 between A, B, and C, so that A gets $\frac{1}{3}$, B $\frac{1}{4}$, and C the balance; what part will C receive?
Ans. C gets $\frac{5}{12}$.

14. Reduce $\frac{22}{33} \times \frac{44}{235} \times \frac{45}{112} \times \frac{1\cdot 4}{4}$ to a simple fraction?   *Ans.* $\frac{1}{4}$.

15. Reduce $\dfrac{2\frac{1}{2} + 3\frac{1}{3}}{3\frac{1}{3} + 2\frac{1}{2}} \div \dfrac{7\frac{1}{2}}{6\frac{3}{4}}$ to a simple fraction.   *Ans.* $\frac{7\frac{3}{4}}{8}$.

86   *FRACTIONS.*

16. Find the greatest common divisor of 105, 231, and 1001.   *Ans.* 7.

   REM.—First find the greatest common divisor of two numbers, then of the G. C. D. found and of the other number.

17. The product of two numbers diminished by $91\frac{1}{4}$ is $153\frac{1}{4}$; one of the numbers is $17\frac{1}{4}$; what is the other number?   *Ans.* $14\frac{5}{23}$.

18. A farmer sold $\frac{2}{3}$ of his cows and $\frac{3}{4}$ of his sheep, and then had 22 cows and 27 sheep left: how many had he at first?   *Ans.* 66 cows and 108 sheep.

19. If $7\frac{1}{4}$ lbs. of sugar cost $71\frac{1}{4}$ cents, what is the cost of $105\frac{1}{4}$ lbs.?   *Ans.* $1002\frac{1}{4}$ cents.

20. What will $8\frac{3}{4}$ barrels flour cost at $\$6\frac{4}{7}$ per barrel ?
          Ans. $\$57\frac{1}{2}$.

21. What cost $136\frac{2}{3}$ lb. ham at $12\frac{1}{2}$ ct. per lb.?
          Ans. $1708\frac{1}{3}$ cents.

22. What cost $180\frac{3}{5}$ tons coal at $\$5\frac{1}{2}$ per ton?
          Ans. $\$992\frac{1}{4}$.

23. What cost $57\frac{1}{4}$ lb. flour at $4\frac{1}{2}$ ct. per lb.?
          Ans. $258\frac{1}{3}$ cents.

24. A and B together have $372, and $\frac{3}{4}$ of A's money is equal to $\frac{4}{5}$ of B's; how much has each?
          Ans. A $192, B $180.

25. If $\frac{1}{7}$ of a ton of coal cost $1, what will a ton cost?
          Ans. $7.

26. If $\frac{2}{7}$ of a ton of coal cost $2, what will a ton cost?
          Ans. $7.

27. If $\frac{1}{6}$ of a cord of wood cost $\$\frac{4}{5}$, what will a cord cost?   *Ans.* $\$4\frac{4}{5}$.

28. If $\frac{5}{6}$ of a cord of wood cost $5, what will a cord cost?   *Ans.* $6.

FRACTIONS.

Cor.—When a given number of articles of the same value costs a certain sum of money, in order to find the cost of one article, divide the certain sum by the given number, whether integral or fractional; and in order to get the cost of any required number, multiply the cost of one article by the required number.

29. If one acre of land is worth $37\frac{1}{2}$, what is the cost of $125\frac{1}{4}$ acres? *Ans.* $4696\frac{7}{8}$.

30. If $\frac{2}{3}$ of a farm cost $4250, what did the whole farm cost? *Ans.* $6375.

31. If $\frac{7}{9}$ of a house is worth $4949, what is $\frac{1}{9}$ of it worth? What is the whole house worth?
Ans. $\frac{1}{9}$ is worth $707, and the house $6363.

32. A watch and chain are worth $150, and the chain is worth $\frac{3}{7}$ the price of the watch; what is the value of each? *Ans.* Watch $105, chain $45.

33. Two-thirds the value of the house is equal to $\frac{3}{4}$ the value of the lot; both are worth $8500; what is the value of each? *Ans.* House $4500, lot $4000.

34. One-third of A's money is equal to $\frac{3}{7}$ of B's; both have $160; how many has each?
Ans. A has $90 and B $70.

35. If $\frac{2}{3}$ of a yard of cloth cost $2\frac{1}{4}$, how many yards can be bought for $143\frac{1}{4}$? *Ans.* $34\frac{4}{11}$ yd.

36. If a man travel $3\frac{3}{4}$ miles an hour, how long will it take him to travel $48\frac{3}{4}$ miles? *Ans.* 13 hours.

37. A cistern has two pipes of supply and one for discharge; the first alone will fill it in 5 hours, the second in 8 hours, and the third will empty it in 9 hours; in what time will it be filled if all run together? *Ans.* $4\frac{44}{49}$ hours.

38. What number is that from which if ⅔ of ¾ of itself be subtracted, the remainder will be 25?

Ans. 50.

39. A father bequeathed to his eldest son $12000, which was ⅗ of his estate; what was the value of his estate? *Ans.* $20000.

40. A, B, and C owned a section of land (640 acres); A owned ⅘ of it, which was ⅞ of what B owned, and C the remainder; how many acres did each own?

Ans. A owned 256 A., B 288 A., and C 96 A.

41. The part of a pole above the water was ⅝ of the part under the water; what part of the pole was above the water and what part under the water?

Ans. ⅝ above and ⅜ under.

42. Find the sum, difference, and product of ⅚ and ⅔; also the quotient, making the larger fraction the divisor? *Ans.* Sum 1½, diff. ⅟₁₂, prod. ⅚, and quo. ⅘.

43. What number must be divided by ⅓ of 36⅔ to produce 4 times 6¼? *Ans.* 183¼.

44. What is the least common multiple of 7½ and 9⅗?

Ans. 435.

45. What is the greatest common divisor of 24¼ and 27⅗? *Ans.* ₃⁄₂₀.

46. A farmer has three separate parcels of land which he wishes to divide into lots of equal size and as large as possible; the first parcel has 5¼ acres, the 2d 4¾ acres, and the 3d 2¼ acres. What will be the size of each lot?

Ans. ¼ acre.

47. A can mow 3 acres of grass in 4 days, and B can mow 4 acres of grass in 5 days; how long will it take both to mow 12 acres? *Ans.* 7²³⁄₃₁ days.

FRACTIONS. 89

48. A and B together can perform a piece of work in 9 days; A alone can do it in 15 days? How long would it take B to do it? *Ans.* $22\frac{1}{2}$ days.

49. A and B can do a piece of work in 5 days, A and C in 6 days, and B and C in 8 days; in what time will each one alone do it?
Ans. A in $8\frac{8}{25}$ days, B in $12\frac{12}{13}$ days, and C in $21\frac{9}{11}$ days.

	A and B	$= \frac{1}{5}$ in a day;
	A and C	$= \frac{1}{6}$ " "
Add,	B and C	$= \frac{1}{8}$ " "
Divide by 2,	$2A + 2B + 2C = \frac{49}{120}$
	$A + B + C = \frac{59}{240}$
Subtract,	$A + B = \frac{48}{240}$
	$C = \frac{11}{240}$

$$A + C = \frac{40}{240}$$
Subtract, $\quad\quad\quad C = \frac{11}{240}$
$$A = \frac{29}{240}$$

$$A + B + C = \frac{59}{240}$$
Subtract, $\quad A + C = \frac{40}{240}$
$$B = \frac{19}{240}$$

50. A merchant bought goods at $\frac{1}{5}$ less than cost, and sold them at first cost; what part of the money he paid was his profit? *Ans.* $\frac{1}{4}$.

51. A and B bought 1000 acres of land, of which A is to have 11 acres as often as B 9 acres; how much does each get? *Ans.* A 550 acres, B 450 acres.

52. A merchant sold $\frac{4}{5}$ of his stock for what the whole cost; what was the rate of profit? *Ans.* $\frac{1}{4}$.

FRACTIONS.

53. A, B, and C engage in business; A puts in $9, to B $8 and C $7; what is each one's part of the stock?
Ans. A $\frac{3}{8}$, B $\frac{1}{3}$, C $\frac{7}{24}$.

54. The profits of a firm this year is 4875\frac{1}{2}$, which is one-fourth more than last year; what were they last year? Ans. 3900\frac{2}{5}$.

55. If $\frac{4}{5}$ of a cord of wood cost $5, what will be the cost of 25$\frac{4}{5}$ cords? Ans. $155.

56. If 25$\frac{4}{5}$ cords of wood cost $155, what will $\frac{2}{3}$ of a cord cost? Ans. $4.

57. Reduce $\frac{11}{2} \times \frac{12}{3} \times \frac{13}{4} \times \frac{14}{5} \div \frac{11}{6} \times \frac{14}{7} \times \frac{13}{8} \times \frac{16}{9}$ to a simple fraction. Ans. $\frac{121}{144}$.

58. Reduce $\dfrac{\frac{1}{2} \times \frac{2}{3} \times \frac{3}{4}}{\frac{4}{5} \times \frac{5}{6} \times \frac{6}{7}}$ to a simple fraction.

Ans. $\frac{7}{16}$.

59. Reduce $\frac{2}{3} \div \frac{7}{8}$ to a simple fraction. Ans. $\frac{2}{3}$.

60. What is the value of $8 \times 12 \times 16 \times 18 \times 24 \div 4 \times 6 \times 32 \times 9 \times 48$? Ans. 2.

61. The difference between $\frac{4}{5}$ and $\frac{2}{3}$ of a number is 21; what is the number? Ans. 90.

62. $\frac{4}{5}$ is $\frac{2}{3}$ of what number? Ans. 1$\frac{1}{5}$.

63. If a pole 6$\frac{1}{2}$ ft. long cast a shadow 9$\frac{3}{4}$ ft., what is the height of a tree that casts a shadow 90$\frac{3}{8}$ ft.? Ans. 60$\frac{3}{8}$ ft.

64. What is the least common multiple of $\frac{1}{2}$, $\frac{2}{3}$, $\frac{3}{4}$, and $\frac{5}{6}$? Ans. 30.

65. What is the greatest common divisor of $\frac{2}{3}$, $\frac{3}{4}$, $\frac{4}{5}$, and $\frac{7}{8}$? Ans. $\frac{1}{84}$.

66. A and B bought goods for $1200; A paid $\frac{3}{5}$ as much as B; what did each pay? Ans. A $450, B $750.

67. Divide $12000 among 4 persons; the first to have $\frac{1}{4}$ as often as the second $\frac{1}{3}$, the third $\frac{1}{4}$, and the fourth $\frac{1}{5}$; what part does each get?

$$\tfrac{1}{2}, \tfrac{1}{3}, \tfrac{1}{4}, \tfrac{1}{5} = \tfrac{30}{60}, \tfrac{20}{60}, \tfrac{15}{60}, \tfrac{12}{60};$$
$$30 + 20 + 15 + 12 = 77.$$

The first to have $\frac{30}{77}$, the second $\frac{20}{77}$, the third $\frac{15}{77}$, and the fourth $\frac{12}{77}$.

68. A, B, and C traded together. A put in $6¼ as often as B $8⅓ and C $9½; the amount of their stock was $2890; what was each one's share?
    Ans. A $750, B $1000, C $1140.

69. The sum of two fractions is $\frac{11}{15}$, their difference is $\frac{3}{15}$; what are the fractions?   *Ans.* $\frac{2}{3}$ and $\frac{1}{3}$.

70. Multiply 37 by 25.
$$37 \times \tfrac{100}{4} = \tfrac{3700}{4} = 925$$

71. Multiply 37 by 2¼.
$$37 \times \tfrac{9}{4} = \tfrac{333}{4} = 92¼$$

72. Multiply 37 by 250.
$$37 \times \tfrac{1000}{4} = \tfrac{37000}{4} = 9250$$

73. Multiply 49 by 225.
$$49 \times \tfrac{900}{4} = 11025.$$

74. Multiply 49 by 333⅓.
$$49 \times \tfrac{1000}{3} = 16333⅓.$$

75. Multiply 49 by 275.
$$49 \times \tfrac{1100}{4} = 13475.$$

DECIMAL FRACTIONS.

ORAL QUESTIONS.

1. In Addition and Subtraction, how do you place the figures as regards the orders?

Ans. Like orders must be placed in the same column; that is, directly under each other.

2. Does the relation of the orders in decimals correspond to that of abstract numbers?

Ans. The relation of any two orders in the same direction is the same.

3. How much greater is the value of the same figure in each consecutive column or order going to the left?

Ans. Ten times.

4. Do the names of the orders taken in the same direction correspond?

Ans. No, but in opposite directions.

5. From what order, going in opposite directions, do the names of the other orders correspond?

Ans. From the order of units, going to the left in the Integral numbers, and to the right in the decimals.

6. What are Decimal Fractions?

Ans. Fractions whose denominators are 10, 100, 1000, etc.

7. Express the decimal fraction in the common form; thus, $\frac{1}{10}$, $\frac{1}{100}$, etc., and decimally, .1, .01, etc.; and these expressions are respectively of the same value.

8. Which do you generally use? Why?

Ans. The common fraction. Because it is more easily understood than the decimal, and is so convenient for cancellation.

DECIMAL FRACTIONS. 93

9. How does the number of decimals in a product correspond to that of the factors?

Ans. The number is equal.

10. How does the number of decimals in a dividend correspond to that of the divisor and quotient?

Ans. It is equal.

11. If the divisor has more decimals than the dividend, what is necessary to be done?

Ans. Add decimal zeros to the dividend until it has as many decimals as the divisor.

12. If the divisor and dividend have the same number of decimals, will there be any in the quotient?

Ans. No.

13. If it be necessary that there should be decimals in the quotient, what must be done?

Ans. More decimal zeros must be added to the dividend.

14. Do zeros annexed to a decimal change its value?

Ans. No.

15. If zeros are prefixed to a decimal, and the decimal point removed to the left of the zeros, is the value changed?

Ans. Each decimal zero thus prefixed renders the value one-tenth, or divides it by ten.

16. In multiplying by more than one figure, where is the first figure of each line or partial product placed, and why?

Ans. It is placed in the column of the same order as the multiplier; that is, directly under it, because every consecutive order to the left is ten times the value of the preceding order.

DENOMINATE NUMBERS.

All arithmetical numbers may be considered Denominate, even abstract numbers, as every figure in each successive order, beginning at the right and going to the left, is ten times the value of the same figure in the previous order, and may be arranged in a table; thus,

    10 units   = 1 ten.
    10 tens   = 1 hundred.
    10 hundred = 1 thousand.
    10 thousand = 1 ten-thousand.

In the United States currency, the orders have the same relation; thus,

    10 mills (*m.*) = 1 cent (*ct.*).
    10 cents   = 1 dime.
    10 dimes   = 1 dollar ($).
    10 dollars = 1 eagle.

Dimes and eagles are coins, but are not regarded in computation; but only dollars ($), cents, and mills, the cents holding two places.

There is generally a decimal point placed between dollars and cents; thus, $456.295, which is numerated "four hundred and fifty-six dollars, twenty-nine cents and five mills. It may also be numerated without any change in its value, "four hundred and fifty-six thousand, two hundred and ninety-five mills.

DENOMINATE NUMBERS. 95

ADDITION.

As the relations of the orders in United States money is the same as in abstract numbers, hence their application is the same; and in addition and subtraction like orders must be placed under each other, and in every other way the same methods are followed.

PROBLEMS.

$25.365	1. What is the sum of twenty-five
12.184	dollars, thirty-six cents and five mills;
9.100	twelve dollars, eighteen cents and four
30.005	mills; nine dollars and ten cents; thirty
15.030	dollars and five mills; fifteen dollars
$91.684	and three cents.

Ans., Ninety-one dollars, sixty-eight cents and four mills.

2. Add the following sums of money:

Five dollars, thirty cents and four mills.	$5.304
Three dollars and two mills	3.002
Two dollars and three cents	2.030
Seven dollars and three mills	7.003
Twelve dollars and one cent	12.010
Nine dollars	9.000
	$38.349

DENOMINATE NUMBERS.

	(3.)		(4.)		(5.)
Add	$97.548	Add	$386.946	Add	387,642 mills.
	68.754		5372.875		548,753
	97.632		64759.654		659,864
	198.564		876943.687		3,217,634
	$462.498		$947463.162		4,813,893 mills.

REM. 1.—The sum of the last example may be numerated thus: Four millions eight hundred and thirteen thousand, eight hundred and ninety-three mills; or, thus: four thousand eight hundred and thirteen dollars, eighty-nine cents and three mills.

REM. 2.—Mills are numerated the same as abstract numbers.

SUBTRACTION.

$287.304
194.293
―――
$93.011

1. From two hundred and eighty-seven dollars, thirty cents and four mills, take one hundred and ninety-four dollars, twenty-nine cents and three mills. Remainder, Ninety-three dollars, one cent and one mill.

(2.)	(3.)	(4.)
$475648.364	$9,486,397.213	$21795.375
387654.875	6,397,423.875	10963.625
$87993.489	$3,088,973.338	$10831.750

(5.)	(6.)	(7.)	(8.)	(9.)
100000	100	100	100.00	100.00
99999	99	1	1.50	2.50
1	1	99	98.50	97.50

REM.—As in addition and subtraction, so also in multiplication, the process is the same as that of abstract integers and decimals; hence there is no need of further exemplification.

DENOMINATE NUMBERS.

English money is reckoned in pounds, shillings, pence, and farthings; sometimes also in guineas; thus,

TABLE.

4 farthings (*far.*) = 1 penny (*d.*).
12 pence   = 1 shilling (*s.*).
20 shillings   = 1 pound (£).
21 shillings   = 1 guinea.

PROBLEMS.

Reduce £1 to shillings, pence, and farthings.

```
  £1
   20                  £1 =  20 shillings.
  ───  shillings.      £1 = 240 pence.
   20                  £1 = 960 farthings.
   12
  ───
  240   pence.
    4
  ───
  960   farthings.
```

As there are twenty shillings in one pound, there will always be twenty times as many shillings as pounds; and as there are twelve pence in every shilling, there will be twelve times as many pence as shillings; and four times as many farthings as pence.

COR.—A higher denomination is reduced to a lower one by multiplication.

Reduce 960 farthings to pence, shillings and pounds; thus,

```
        4 ) 960 farthings.
       12 ) 240 pence.
       20 )  20 shillings.
              1 pound.
```

As four farthings make one penny, there will be one-fourth as many pence as farthings, one-twelfth as many shillings as pence, and one-twentieth as many pounds as shillings.

Cor.—A lower denomination is reduced to a higher one by division.

Reduce 1095 farthings to pence, shillings, and pounds.

```
    4 ) 1095 farthings.
   12 )  273 . . . 3 far.
   20 )   22 . . . 9d.
          £1 2s. 9d. 3 far.
```

The first remainder is farthings, the second pence, and the third shillings.

Reduce   £1   2s.   9d.   3 far.   to farthings.

```
           20
           ――
           22 shillings.
           12
           ――
          273 pence.
            4
           ――
         1095 farthings.
```

In reducing a higher denomination to a lower one, begin by multiplying by the number of the next lower denomination that makes one of the higher, and if it be a compound number, add to the product the number of the lower denomination, and continue this process until you reach the lowest denomination.

In reducing a lower to a higher denomination, divide by the number of the lowest denomination that makes one of the next higher, and if there be a remainder, it will be of the lowest denomination, etc.

DENOMINATE NUMBERS.

Cor.—In the computation of compound numbers, instead of carrying a unit to a higher order for every ten, as in abstract numbers, a unit is carried to a higher denomination as often as the sum reaches the number that it takes of the lower denomination to make one of the next higher denomination; thus, as 4 farthings make 1 penny, as often as the sum of the farthings reaches four, one must be carried to the pence; and as 12 pence make 1 shilling, in computing pence as many must be carried to shillings as the number of times 12 is contained in the number of pence; 1 from shillings to pounds for every 20.

In division, the order is reversed, as then we begin with the highest denomination and descend.

EXAMPLES.

	£	s.	d.	far.
1. Add	3	8	7	3
	5	9	6	2
	6	11	9	1
	8	15	11	3
	24	5	11	1

The sum of the first column is 9 farthings, which is 2 times 4 and 1; the 1 is farthings, and must be placed under the farthings; the 2 is carried to the next denomination and added with the pence, the sum of which is 35; that is, 2 times 12 and 11, that is, 2 shillings and 11 pence; the 2 is added with the shillings, making the sum 45, which is £2 5s.; the shillings are placed under the shillings and the 2 carried to the pounds, the sum of which is 24.

100   DENOMINATE NUMBERS.

	£	s.	d.	far.
2. From	54	6	5	1
	28	7	6	3
	£25	18s.	10d.	2 far.

As you cannot subtract 3 farthings from 1 farthing, you must borrow 1 penny, which is 4 farthings; this 4 and the 1 make 5; then 3 from 5, 2 remains; the 1 penny borrowed must be carried to the 6, which makes 7, which cannot be subtracted from 5; 1 shilling, that is, 12 pence, must be borrowed and added to the 5, which makes 17; 7 from 17, 10 remains; 1 shilling to carry to 7 makes 8, which cannot be taken from 6; 1 pound, that is, 20 shillings, must be borrowed and added to the 6, making 26, from which subtract 8 and 18 remains; and £1 to carry to 28, making 29, which is subtracted from 54 and 25 remains.

Rem.—When the subtrahend is less than the minuend, the difference can be taken directly.

	£	s.	d.	far.
3. Multiply	4	6	5	3
by				5
	£21	12s.	4	3

```
   4         6          5 .        3
   5         5          5          5
  ──       ────        ──        ────
  21      20)32        25       4)15
           £1 12s.      3        3d. 3 far.
                    12)28
                     2s. 4d.
```

Cor.—Multiply each denominate number, and divide the product by the number of this denomination that it takes to make one of the higher, and carry the number of times it is contained to the higher denomination, and place the remainder under its kind.

4. Multiply £48 12s. 7d. 2 far. by 6.
5. Divide 4) £5 6s. 3d. 1 far. by 4.
     £1 6s. 6d. 3¼ far.

4 is contained in 5, once and £1 over; this £1 is 20 shillings, which added to the 6 shillings make 26 shillings, into which 4 is contained 6 times and 2 shillings over; this 2 shillings is 24 pence, which added to the 3 pence, makes 27 pence, in which 4 is contained 6 times and 3 pence over, which is 12 farthings, and 1 more make 13, in which 4 is contained 3¼ times.

6. Divide £754 15s. 9d. 3 far. by 27.

  27) £754 15s. 9d. 3 far. (£27
     54
    ―――
     214
     189
    ―――
      25
      20
    ―――
     515 (19s. Add the 15s.
      27
    ―――
     245
     243
    ―――
       2
      12
    ―――
      33 (1d. Add the 9d.
      27
    ―――
       6
       4
    ―――
      27 (1 far. Add the 3 far.
      27
    ―――

Quotient = £27 19s. 1d. 1 far.

DENOMINATE NUMBERS.

7. Multiply   £5   4s.   6d.   1 far.   by 35.
                           35
   ─────────────────────────────
          £182 18s. 2d. 3 far.

```
   35         35          35       4)35
    5          4           6       ─────
   ───        ───         ───      8d. 3 far.
   175        140         210
     7         18           8
   ───     20)158(7     12)218
   £182       140           18s. 2d.
               18
```

REM.—Observe these solutions carefully; for if they are understood, there is no further difficulty in denominate numbers; the principle is the same in all, the tables alone differ.

EXAMPLES

1. In 2 dollars, how many cents? How many mills?

$$\$2 \times 100 = 200 \text{ cents.}$$
$$2 \times 1000 = 2000 \text{ mills.}$$

2. In 5 dollars, how many cents? How many mills?
3. In 7 dollars, how many cents? How many mills?
4. In 5 dollars 15 cents, how many cents? How many mills?

$$\$5 = 500 \text{ cents.}$$
$$15$$
$$\overline{515 \text{ cents} = 5150 \text{ mills.}}$$

5. In 6 dollars 15 cents and 3 mills, how many mills?
6. In 500 cents, how many dollars? $\frac{500}{100} = \$5$, *Ans.*
7. In 625 cents, how many dollars and cents?

Ans. $6.25.

DENOMINATE NUMBERS.   103

8. In 5325 mills, how many dollars, cents, and mills?
   Ans. $5.325.

9. In 63257 mills, how many dollars, cents, and mills?

10. In 75325 cents, how many dollars and cents?

11. If 1 bushel of wheat cost $1.125, what will 8 bushels cost?

12. If 1 bushel of wheat cost $1.05, what will 10 bushels cost?

13. If 1 bushel of wheat cost $1.05, what will 100 bushels cost?

14. If 8 bushels of wheat cost $9, what cost 1 bushel?

15. If 8 bushels of wheat cost $9, what cost 35 bushels?

16. If 10 bushels of wheat cost $10.50, what cost 53 bushels?

17. Bought dry goods for $243.37; groceries for $146.294; hardware for $71.96; notions for $21.512. What was the amount of the bill? Sold the same at a profit of $157.192. What did I sell the whole for?

18. If 5 lbs. sugar cost 50 cents, what will 6 lbs. cost? 7 lbs.? 8? 9? 10? 11? 12?

19. If 6 lbs. cost 72 cts., what will 7 lbs. cost? 8 lbs.? 9? 10? 11? 12?

20. In 15 farthings, how many pence?
   Ans. 3¾ pen

21. In 18 farthings, how many pence? How many pence in 21 far.? 23? 25? 27? 29? 31? 33? 34? 35?

22. How many shillings in 25 pence? in 28? 35? 38? 45? 51? 56? 65?

23. How many pounds in 35 shillings? in 40? 50? 60? 65? 70? 75? 80? 85? 90? 95? 100? 105? 110? 120?

DENOMINATE NUMBERS.

24. How many farthings in £9 13s. 9d. 3 far.?
25. How many pounds, shillings, pence, and farthings in 37864321 farthings?
26. Multiply £4 8s. 9d. 3 far. by 9.
27. Divide £25 9s. 4d. 1 far. by 13?

AVOIRDUPOIS, or COMMERCIAL WEIGHT,

is used in commercial transactions, when goods are bought or sold in quantity, and for all metals except gold and silver.

TABLE.

16 drams (*dr.*) = 1 ounce (*oz.*)
16 ounces = 1 pound (*lb.*).
25 pounds = 1 quarter (*qr.*).
4 quarters = 1 hundredweight (*cwt.*).
20 cwt. = 1 ton (*T.*).

EXEMPLIFICATION.

1 T.
20
―――
20 cwt.
4
―――
80 qrs.
25
―――
2000 lbs.
16
―――
32000 oz.
16
―――
512000 dr.

16) 512000 dr.
―――――――――
16) 32000 oz.
―――――――――
25) 2000 lbs.
―――――――――
4) 80 qrs.
―――――――――
20) 20 cwt.
―――――――――
1 T.

DENOMINATE NUMBERS.

```
         T. cwt. qr. lb. oz. dr.
Reduce   3   4   2   8   6   10
        20
        ──
        64
         4
        ───
        258                 16 ) 1653354
         25                 16 ) 103334 ... 10 dr.
        ────                25 ) 6458 ... 6 oz.
        6458                 4 ) 258 ... 8 lb.
          16                20 ) 64 ... 2 qr.
        ─────
        103334                   3   4   2   8   6   10
          16                    T. cwt. qr. lb. oz. dr.
        ──────
        1653354 drams.
```

Reduce 1653354 drams to the original denominations.

TROY WEIGHT

is used for gold, silver, and jewels; also in philosophical experiments.

TABLE.

24 grains (*gr.*) = 1 pennyweight (*pwt.*).
20 pennyweights = 1 ounce.
12 ounces = 1 pound.

```
     1 lb.
      12                  24 ) 5760 gr.
     ────                 20 ) 240 pwt.
     12 oz.               12 ) 12 oz.
      20                          1 lb.
     ────
     240 pwt.
      24
     ────
     5760 gr.
```

Reduce   5 lb.   6 oz.   10 pwt.   16 gr.
         12
         ──
         66   Add the 6 oz.
         20
         ────
         1330   Add the 10 pwt.
         24
         ─────
         31936   Add the 16 gr.

Reduce 31936 grs. to the original denominations.

24) 31936 gr.
20) 1330 ... 16 gr.
12) 66 ... 10 pwt.
       5 lb. 6 oz. 10 pwt. 16 gr.

DIAMOND WEIGHT.

Used for diamonds and other precious stones.

TABLE.
16 parts = 1 grain = .8 grain Troy.
4 grains = 1 carat = 3.2 grains Troy.

APOTHECARIES' WEIGHT

is used by druggists in putting up prescriptions; the pound, ounce, and grain are the same as in Troy Weight.

TABLE.
20 grains   = 1 scruple (℈).
3 scruples = 1 dram (ʒ).
8 drams   = 1 ounce (℥).
12 ounces   = 1 pound.

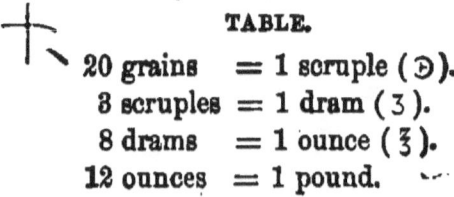

DENOMINATE NUMBERS. 107

```
    1 lb.
    12
   ─── ʒ              20 ) 5760 gr.
    12
    8                  3 )  288  ℈
   ─── ʒ
    96                 8 )   96  ʒ
    3
   ─── ℈              12 )   12  ʒ
   288                       1 lb.
    20
  ─────
  5760 gr.
```

APOTHECARIES FLUID WEIGHT

is used for liquids in medical prescriptions.

TABLE.

60 minims (♏︎) = 1 fluid dram (f ʒ).
8 fluid drams = 1 fluid ounce (f ℥).
16 fluid ounces = 1 pint (O.).
8 pints = 1 gallon (*Cong.*).

For ordinary use, 1 teacup = 2 wine glasses = 8 table-spoons = 32 tea-spoons = 4 f ℥.

COMPARISON OF WEIGHTS.

1 lb. Avoirdupois = 7000 gr. Troy.
1 lb. Troy = 5760 gr. Troy.

LINEAR MEASURE

is used for lengths and distances.

TABLE.

12 inches (*in.*) = 1 foot (*ft.*).
3 feet = 1 yard (*yd.*).
5½ yds., or 16½ ft. = 1 rod (*rd.*).
40 rods = 1 furlong (*fur.*).
8 furlongs = 1 mile (*m.*).
3 miles = 1 league (*lea.*).

MARINER'S MEASURE.

6 feet	= 1 fathom.
120 fathoms	= 1 cable length.
880 fath., or 7½ cable lengths	= 1 mile.

REM.—1 nautical league = 3 equatorial miles = 3.45771 statute miles. 60 equatorial miles = 69.1542 statute miles = 1 equatorial degree (°). 360° = the circumference of a circle. 360 equatorial degrees = the circumference of the earth.

CLOTH MEASURE.

2¼ inches (*in.*)	= 1 nail (*na.*).
4 nails or 9 in.	= 1 quarter.
4 quarters	= 1 yard.
3 quarters	= 1 ell Flemish.
5 quarters	= 1 ell English.
6 quarters	= 1 ell French.

SURVEYOR'S MEASURE OF LENGTH.

7$\frac{92}{100}$	inches	= 1 link (*l.*).
25	links	= 1 pole (*p.*).
100	links, 4 poles, 66 feet	= 1 chain (*ch.*).
10	chains	= 1 furlong.
8	furlongs, or 80 chains	= 1 mile.

LAND MEASURE.

40 perches	= 1 rood.
4 roods	= 1 acre.
640 acres	= 1 square mile, termed a Section.

SQUARE MEASURE.

A *Square* is a surface bounded by four equal sides, its angles are also equal; thus,

This figure, ABCD, represents a square foot, considering each of the small spaces as an inch. The sides, AB, BC, AD, and DC, each equal to 12 inches in length. The angles, A, B, C, and D, are equal; that is, if one is placed on the other, the sides respectively will coincide.

AB is 12 inches long, and every inch in width makes 12 square inches; and the 12 inches in width, which is either AD or BC, makes $12 \times 12 = 144$ sq. in.; hence the correspondence of linear and square measure; thus,

LINEAR MEASURE.

12 inches = 1 foot.
3 feet = 1 yard.
5½ yards = 1 rod, pole, or perch.

SQUARE MEASURE.

$12 \times 12 = 144$ square inches = 1 square foot.
$3 \times 3 = 9$ square feet = 1 square yard.
$5\frac{1}{2} \times 5\frac{1}{2} = 30\frac{1}{4}$ square yards = 1 square rod; also called perch.

CUBIC MEASURE.

$12 \times 12 \times 12 = 1728$ cubic inches = 1 cubic foot.
$3 \times 3 \times 3 = 27$ cubic feet = 1 cubic yard.

CUBIC MEASURE.

A ***Cube*** is a solid figure bounded by six equal squares; the square on page 72 represents the base or any other side, as the sides are all equal; the length of all the edges are equal, and the angles are all equal. If the above figure have 12 inches altitude added to it, every inch will make 144 cubic inches and 12 inches in altitude, 144 × 12 = 1728, which is the number of cubic inches in a cubic foot.

TABLE.

1728 cubic inches	= 1 cubic foot.
27 cubic feet	= 1 cubic yard.
16 cubic feet	= 1 cord foot.
128 cubic feet, or 8 cord feet	= 1 cd. of wood, bark, etc.

40 cubic feet of round timber, or 50 cubic feet of hewn timber = 1 Ton.

A perch of stone is $16\frac{1}{2}$ feet long, $1\frac{1}{2}$ feet wide, and 1 ft. high = $24\frac{3}{4}$ solid feet.

LIQUID MEASURE.

This measure is used for all liquids.

TABLE.

4 gills (*gi.*)	= 1 pint (*pt.*).
2 pints	= 1 quart (*qt.*).
4 quarts	= 1 gallon (*gal.*).

In all liquids, except ale, beer, and milk, the gallon is 231 cubic inches.

In ale, beer, and milk, it is 282 cubic inches.

DENOMINATE NUMBERS.

Rem.—In the former 31½ gallons is called a barrel, 63 gallons a hogshead, 42 gallons a tierce, 84 gallons a puncheon, and 126 gallons a pipe, and 2 pipes a tun. In the latter, 36 gallons = a barrel, and 54 gallons = a hogshead; these, however, are not measures, but only vessels.

DRY MEASURE

is used for grain, fruits, vegetables, coal, salt, etc.

TABLE.

2 pints = 1 quart.
8 quarts = 1 peck ($pk.$).
4 pecks = 1 bushel ($bu.$)
= 2150.42 cubic inches.

The wine gallon of United States = 231 cu. in.
The beer gallon of United States = 282 cu. in.
The dry gallon of United States = 268.8 cu. in.
Imperial gal. of Great Britain for dry
 and liquid measures = 277.274 cu. in.
Dry bushel of United States = 2150.42 cu. in.
Imperial bushel of Great Britain = 2218.192 cu. in.

TIME TABLE.

60 seconds ($sec.$) = 1 minute ($m.$).
60 minutes = 1 hour ($hr.$).
24 hours = 1 day ($da.$).
7 days = 1 week ($wk.$).
30 days = 1 month ($mo.$).
365 days = 1 common year.
366 days = 1 leap year.

ANGULAR OR CIRCULAR MEASURE

is applied to angles and circumferences, reckoning latitudes and longitudes, etc.

TABLE.

60 seconds (") = 1 minute (').
60 minutes = 1 degree (°).
30 degrees = 1 sign (S.).
12 signs or 360 degrees = 1 circumference.

Apparently the sun makes an entire revolution of the earth in 24 hours,* and consequently travels 15° in 1 hour; therefore,

1 hour of time = 15° longitude.
1 minute of time = 15' longitude.
1 second of time = 15" longitude.

MISCELLANEOUS TABLE.

12 units = 1 dozen.
12 dozen = 1 gross.
12 gross = 1 great gross.
20 units = 1 score.
24 sheets of paper = 1 quire.
20 quires = 1 ream.
196 lbs. = 1 barrel of flour.
200 lbs. = 1 barrel of pork.

When a sheet of paper is folded into two leaves, or 4 pages, and a book made in this way, it is called a folio.

4 leaves is called a quarto.
8 leaves is called an octavo.
12 leaves is called a duodecimo.

* *Really*, the revolution is that of the earth on its own axis.

DENOMINATE NUMBERS. 113

PRACTICAL QUESTIONS.

1. Reduce £3 9s. 11d. 3 far. to farthings.
2. Reduce £12 15s. 8d. to pence.
3. Reduce £7 0s. 2d. to pence.
4. Reduce 2354 farthings to the higher denominations.
5. Reduce 543 pence to the higher denominations.
6. Reduce 731 shillings to the higher denominations.
7. Reduce 3 T. 6 cwt. 2 qr. 12 lb. 6 oz. and 9 dr. to drams.
8. Reduce 672432 drams to the higher denominations.
9. Reduce 5 lb. 8 oz. 9 pwt. 15 gr. to grains.
10. Reduce 64324 grains Troy to the higher denominations.
11. Reduce 2 lb. 6 ℥ 4 ʒ 2 ℈ 10 gr. to grains.
12. Reduce 6742 gr., Apothecaries weight, to the higher denominations.
13. Reduce 3 lea. 2 mi. 5 fur. 24 rods 2 yd. 1 ft. 6 in. to inches.
14. Reduce 802456 inches to the higher denominations.
15. Reduce 4 yards 3 qrs. 2 na. and 2 inches to inches.
16. Reduce 5 ells Flemish to inches.
17. Reduce 4 ells English to inches.
18. Reduce 3 ells French to inches.
19. Reduce 4 sq. rods 8 sq. yd. 105 sq. ft. and 112 sq. in. to square inches.
20. Reduce 3 cu. yd. 12 cu. ft. and 1236 cu. in. to cubic inches.
21. A pile of wood is 16 ft. long, 4 ft. high, $\frac{16 \times 4 \times 4}{8 \times 4 \times 4}$ and the length of the wood is 4 feet. How many cords of wood?

114 DENOMINATE NUMBERS.

22. Reduce 25 gal. 3 qt. 1 pt. and 3 gills to gills.
23. Reduce 9 bu. 3 pk. 4 qt. 1 pt. to pints.
24. How many years, months, and days, from April 15th, 1842, to June 20th, 1850?

yr.	mo.	da.
1850	6	20
1842	4	15
8	2	5

25. How many years, months, and days, from October 25th, 1845, to August 18th, 1850?

yr.	mo.	da.
1850	8	18
1845	10	25
4	9	23

Rem.—In this question, instead of using the eighth and tenth months, some authors prefer calling them 7 months and 9 months; but if there is an inaccuracy in the months, there is also in the years; hence if we read the above thus: The one thousand eight hundred and fiftieth year, the 8th month and 18th day, there is no inaccuracy.

In computations of time we always take 30 days as a month.

26. The difference in time of two places is 2 hr. 2 min. and 2 sec.; what is the difference in longitude?

hr.	min.	sec.
2	2	2
		15
30°	30'	30''

Ans. Thirty degrees, 30 minutes, and 30 seconds.

27. If the difference of longitude of two places is 15°, the difference of time will be one hour; the eastern place will have the latest time. If the difference in longitude is 16° 24′ 30″, what is the difference in time?

15) 16° 24′ 30″ (1 hr. 5 min. 38 sec.

```
15
--
 1
60
--
84 ( 5
75
--
 9
60
---
570 ( 38
45
---
120
120
---
```

10 cts.	= $\$\frac{1}{10}$.
$12\frac{1}{2}$ cts.	= $\$\frac{1}{8}$.
$16\frac{2}{3}$ cts.	= $\$\frac{1}{6}$.
20 cts.	= $\$\frac{1}{5}$.
25 cts.	= $\$\frac{1}{4}$.
$33\frac{1}{3}$ cts.	= $\$\frac{1}{3}$.
$37\frac{1}{2}$ cts.	= $\$\frac{3}{8}$.
50 cts.	= $\$\frac{1}{2}$.
$62\frac{1}{2}$ cts.	= $\$\frac{5}{8}$.
$66\frac{2}{3}$ cts.	= $\$\frac{2}{3}$.
75 cts.	= $\$\frac{3}{4}$.
$87\frac{1}{2}$ cts.	= $\$\frac{7}{8}$.

28. Multiply 576 by 100 = 57600.

29. Multiply 576 by 25 = $\frac{1}{4}$ = 14400. Take $\frac{1}{4}$ of the above.

30. Divide 576 by 100 = 5.76.

31. Divide 576 by 25 = 23.04. Multiply by 4.

32. Multiply 576 by 50 = $\frac{1}{2}$ (57600) = 28800.

33. Divide 576 by 50 = 5.76 × 2 = 11.52.

34. Multiply 576 by $.12\frac{1}{2}$ = $\frac{1}{8}$ = \$72.

35. Multiply 576 by $.16\frac{2}{3}$ = $\frac{1}{6}$ = \$96.

36. Multiply 576 by $.33\frac{1}{3}$ = $\frac{1}{3}$ = \$192.

37. Multiply 576 by $.62\frac{1}{2}$ = 576 × 5 ÷ 8 = 360.

38. Multiply 576 by $.87\frac{1}{2}$ = 576 × 7 ÷ 8 = 504.

39. What cost 342 yds. muslin at $12\frac{1}{2}$ cts. per yard?

8) 342

$\$42\frac{6}{8} = \$42\frac{3}{4}$.

40. What cost 342 yds. linen at 37½ cts. per yard?
Multiply by 3 = $128¼.

41. What cost 342 yds. linen at 62½ cts. per yard?
Multiply by 5 = $213¾.

42. What cost 342 yds. linen at 87½ cts. per yard?
Multiply by 7 = $299¼.

43. What cost 548 yds. muslin at 16⅔ cts. per yard?

6) 548
$91⅓

44. What cost 345 yds. muslin at 20 cts. per yard?

5) 345
$69

45. What cost 469 yds. linen at 33⅓ cts. per yard?

3) 469
$156⅓

46. What cost 469 yds. linen at 66⅔ cts. per yard?
Multiply by 2 = $312⅔.

47. What cost 500 yds. linen at 25 cts. per yard?
500 ÷ 4 = $125.

48. What cost 500 yds. linen at 75 cts. per yard?
Multiply by 3 = $375.

49. What cost 500 yds. linen at 50 cts. per yard?
500 ÷ 2 = $250.

50. Bought 648 yards muslin at 12½ cts. a yard, and sold it at 16⅔ cts. per yard. What was the profit?

⅙ of 648 = $108
⅛ of 648 = 81
$27, Profit.

51. Bought 500 yds. cloth at 20 cts., and sold it at 25 cts.; what profit?

¼ of 500 = $125
⅕ of 500 = 100
$25, Profit.

52. Bought 480 yds. cloth at 66⅔ cts., and sold it at 87½ cts.; what was the profit?

$$\tfrac{1}{8} \text{ of } 480 = \$60; \quad \tfrac{7}{8} = \$420$$
$$\tfrac{1}{3} \text{ of } 480 = 160; \quad \tfrac{2}{3} = \underline{320}$$
$$\$100, \text{ Profit.}$$

53. Bought 480 yds. at 37½ cts., and sold it for 50 cts. per yard; what was the profit?

$$\tfrac{1}{2} \text{ of } 480 = \$240$$
$$\tfrac{1}{8} \text{ of } 480 = 60; \quad \tfrac{3}{8} = \underline{180}$$
$$\$60, \text{ Profit.}$$

54. Bought 600 yds. cloth at $1 per yard, and sold it for $1.25 per yard; what was the profit?

$$600 \times 1\tfrac{1}{4} = \$750$$
$$600 \times 1 = \underline{600}$$
$$\$150, \text{ Profit.}$$

55. How many yards of cloth, at 12½ cts. per yard, can be bought for $240? 8 yds. can be bought for every dollar.

$$240 \times 8 = 1920 \text{ yds.}$$

56. How many yards for 16⅔ cts.? 20 cts.? 25 cts.? 37½ cts.? 50 cts.? 62½ cts.? 75 cts.? 87½ cts.?

$$37\tfrac{1}{2} = \tfrac{3}{8}; \quad \overset{80}{\cancel{240}} \times \tfrac{8}{3} = 640 \text{ yds.};$$
$$\text{for } 62\tfrac{1}{2} \text{ cts.} = 240 \times \tfrac{8}{5}.$$

57. Sold 500 barrels flour at $6.62½ per barrel, and invested the proceeds in different kinds of dry goods, averaging 87½ cts. per yard. What were the proceeds of the flour, and how many yards of goods did I get?

COMPARATIVE VIEW OF DENOMINATE NUMBERS AND FRACTIONS.

Thus, £4   5s.   8d.   3 far.
       20   12   4
       ──   ──   ──
       80   60   32
       12   4
       ──   ───
       960   240   240
        4   32
       ────   8
       3840   ───
                     275
£4 + $\frac{5}{20}$ + $\frac{8}{240}$ + $\frac{3}{960}$. $\frac{275}{960} = \frac{55}{192}$.

hence, £4 5s. 8d. 3 far. = £4$\frac{55}{192}$.

    3840   4 × 960 = 3840
     240   5 × 48 = 240
      32   8 × 4   =   32
       3   3   =   3
    ─────   ─────
    4115 farthings.   4115
                                 ──── = £4$\frac{55}{192}$.
                                  960

The common fraction may be reduced to a decimal; thus, $\frac{55}{192}$ = .2864583.

The denominate numbers are all reduced to the lowest denomination, and the whole number and fractions are all reduced to a common denominator.

COR. 1.—As the number of each denomination may properly be regarded as units, so the number expressing the numerator of any fraction may be regarded as units, and used accordingly.

TIME.

PROBLEM I.

To find the difference of time between two given dates, subtract the former date from the latter.

EXAMPLES.

1. What is the time from the 10th April, 1845, to the 15th July, 1850?

	yr.	mo.	da.
	1850	7	15
	1845	4	10
Ans.	5	3	5

2. What is the time between 15th July, 1850, and the 10th April, 1856?

	yr.	mo.	da.
	1856	4	10
	1850	7	15
Ans.	5	8	25

3. What is the time from the 15th May, 1856, to the 1st January, 1860? *Ans.* 3 yr. 7 mo. 16 da.

4. How many years, months and days from March 10th, 1872, to June 20th, 1874? *Ans.* 2 yr. 3 mo. 10 da.

5. What is the time from Sept. 13th, 1865, to April 10th, 1870? *Ans.* 4 yr. 6 mo. 27 da.

6. What is the time from Dec. 25th, 1870, to Jan. 13th, 1875? How many days? *1st Ans.* 4 yr. 18 da.

$$360 \times 4 = 1440$$
$$18$$

2d Ans. 1458 days.

PROBLEM II.

To find the difference of time of two places whose longitudes are given.

REM.—Divide the difference of longitude by 15, and in the quotient regard degrees, minutes and seconds as hours, minutes and seconds of time.

The time must be rated from the fixed meridian, which for the world is Greenwich, England, and as the sun appears to move west, every place west of this meridian must have earlier time, at the rate of 1 hour for every 15 degrees; that is, when it is 9 o'clock A.M., at Greenwich, it is 8 A.M. 15 degrees west of Greenwich.

EXAMPLES.

1. The longitude of New York is 74° 1' 6" west, and that of Cincinnati is 84° 24' west. What is the time at Cincinnati, when it is 12 o'clock, noon, at New York.

```
         84°  24'   0"
         74    1    6
      _____
   15 ) 10°  22'  54"  ( 0 hr. 41 min. 31⅔ sec.
         60
        _____
        622 ( 41                 12  00   00
         60                          41   31⅔
        ____                    _____
         22              Ans.   11  18   28⅓, A.M.
         15
        ____
          7
         60
        _____
        474 ( 31⅔
         45
        ____
         24
         15
        ____
         9/15 = 3/5
```

2. Longitude of New York is 74° 1' 6" and St. Louis 90° 15' 10". When it is noon at New York, what is the time at St. Louis?

```
          90°   15'   10"
          74     1     6
     15 ) 16°   14'    4"   ( 1 hr. 4 min. 56 4/15 sec.
          15
          ——
           1                  hr.  min.  sec.
          60                  12   00    00
          ——                   1    4    56 4/15
          74  ( 4 min.    Ans. 10   55    3 11/15
          60
          ——
          14
          60
          ——
          844 ( 56 sec.
          75
          ——
          94
          90
          ——
          4/15
```

3. The longitude of Dublin is 6° 20' 30" west, and of Louisville 85° 30' west. What is the time at Louisville, when it is 2 A.M. at Dublin?

```
     85°   30'   00"              hr.   min.   sec.
      6    20    30                2    00     00
     ———————————                    5    16     38
     79°    9'   30 ÷ 15 =    Ans.  8    43     22 P.M.
                                   Of the previous day.
```

PROBLEM III.

To find the difference of longitude of two places, the difference in their time being given.

Multiply the difference of time by 15, and regard the hours, minutes and seconds of the product as degrees, minutes and seconds of longitude.

EXAMPLES.

1. The difference in time of two places is 3 hr. 45 min. 30 sec.; what is the difference in longitude?

hr.	min.	sec.
3	45	30
		15
56°	22′	30″, diff. in long.

2. The difference in time between New York and Cincinnati is 41 min. $31\frac{3}{5}$ sec.; the longitude of New York is 74° 1′ 6″ west. What is the longitude of Cincinnati.

	min.	sec.
	41	$31\frac{3}{5}$
		15
10°	22′	54″
74	1	6
84°	24′, west long. of Cincinnati.	

DENOMINATE NUMBERS. 123

3. The time at St. Louis is 1 hr. 4 min. $56\frac{4}{15}$ sec. earlier than that of New York, whose longitude is 74° 1' 6". What is the longitude of St. Louis?

hr.	min.	sec.
1	4	$56\frac{4}{15}$
		15
16°	14'	4", diff. in long.
74	1	6
90°	15'	10", long. of St. Louis.

4. The time at Louisville is 5 hr. 16 min. 38 sec. earlier than that of Dublin, whose longitude is 6° 20' 30" west. What is the longitude of Louisville?

hr.	min.	sec.
5	16	38
		15
79	9	30
6	20	30
85°	30'	00", long. of Louisville.

REM.—When it is 9 A.M. at New York it is 10 A.M. 15° east of New York and 15° west of New York the time is 8 A.M.

5. The difference in time of Washington and St. Petersburg is 7 hr. 9 min. $19\frac{1}{4}$ sec. What is the difference in longitude?

Ans. 107° 19' $48\frac{3}{4}$" = difference in longitude.

6. The longitude of Cincinnati is 84° 24' west. When it is 10 o'clock A.M. at Cincinnati, it is 22 min. 8 sec. past 10 at Buffalo. What is the longitude of Buffalo?

Ans. Longitude of Buffalo 78° 52'.

PRACTICAL EXAMPLES.

1. What is the cost of 40 bushels of wheat at $1.25 per bushel? *Ans.* $50.

2. What is the cost of 110 bushels of wheat at $1.25 per bushel? *Ans.* $137.50.

3. Sold at one time 57 bushels of wheat at $1.20 per bushel, at another 140 bushels at $1.25 per bushel, at a third sale 436 bushels at $1.30 per bushel; what was the amount of sales? *Ans.* $810.20.

4. I paid $1236.43 on a debt of $2143.24; how much do I still owe? *Ans.* $906.81.

5. Bought 237 acres 1 rod 20 perches of land, at $45.50 per acre; what is the cost?

$237\frac{3}{8}$ A. = 237.375 A. *Ans.* 10800.56\frac{1}{4}$.

6. Purchased 450 acres of land for $15000; what was the price per acre? *Ans.* 33\frac{1}{3}$.

7. John Jones bought of Peter Smith,

  4 doz. handks. at 12$\frac{1}{2}$ cts. each ($\frac{1}{8}$),
  60 yd. muslin at 8$\frac{1}{4}$ cts. per yd.,
  25 yd. drilling at 12$\frac{1}{2}$ cts. per yd. ($\frac{1}{8}$),
  42$\frac{1}{4}$ yd. cloth, at 2.37\frac{1}{2}$ per yd. 2\frac{3}{8}$.

What was the amount of goods? *Ans.* 115.01\frac{1}{4}$.

8. 3 dollars per dozen is how much apiece?
Ans. 25 cents.

DENOMINATE NUMBERS. 125

9. 6 dollars per dozen is how much apiece?
Ans. 50 cents.
10. 9 dollars per dozen is how much apiece?
Ans. 75 cents.
11. 37½ cents each is how much per doz.?
Ans. $4.50.
12. 62½ cents each is how much per doz.?
Ans. $7.50.
13. 87½ cents each is how much per doz.?
Ans. $10.50.
14. Sold 565¼ bushels of wheat for $636.18¾; what was the price per bushel? *Ans.* $1.12¼.
15. Paid $150 for carpeting at $1.25 per yard; how many yards were bought? *Ans.* 120 yards.
16. If 4 lb. sugar cost $.37½, what will 25 lb. cost?
Ans. $2.34⅜.
17. A, B, and C bought 600 acres of land; A is to have .3 of it, B .25, and C the balance; what is C's part, and how many acres does each get?
Ans. C gets .45 = 270 A., A gets 180 A., and B 150 A.
18. Add the following: 5 T. 2 cwt. 30 lb. 12 oz. 10 dr., 8 T. 5 cwt. 63 lb. 8 oz. 9 dr., 15 T. 5 cwt. 75 lb. 5 oz. 6 dr., 12 T. 9 cwt. 41 lb. 5 oz. 6 dr., 14 T. 6 cwt. 80 lb. 5 oz. 3 dr. Put cwt. and lb. together. Thus,

T.	lb.	oz.	dr.
5	230	12	10
8	563	8	9
15	575	5	6
12	941	5	6
14	680	5	3
55	991	5	2,

The denomination of lb. is the most important in Avoirdupois Weight. The oz. may be regarded as a fraction, and then 2000 lb. = 1 Ton is all *Ans.* that is necessary.

19. From 8 T. 563 lb. 8 oz. 9 dr. take 5 T. 230 lb. 12 oz. 10 dr. *Ans.* 3 T. 332 lb. 11 oz. 15 dr.

20. Reduce 5 lb. 8 oz. 12 pwt. 14 gr. to grains.
Ans. 32942 gr.

21. One lb. Troy is what part of a lb. Avoirdupois?
Ans. $\frac{144}{175}$.

22. Reduce 1 mi. 2 fur. 8 rd. 3 yd. 2 ft. 8 in. to inches.
Ans. 80924 in.

23. How many yards in 4 pieces of cloth, each containing 35 yd. 3 qr.? *Ans.* 143 yd.

24. What is the value of 5 T. 420 lb. coal, @ $4¼ per ton? *Ans.* $22.14¼.

25. In 12 lb. 6 oz. 10 pwt. 12 gr., reduce the lower denominations to the fraction of a lb.
Ans. $12\frac{3483}{6400} = 12\frac{87}{160}$, or 12.54375.

26. Reduce the lower denominations to the fraction of an acre in 30 A. 3 R. 12 P.
Ans. 30.825 A., or $30\frac{33}{40}$ A.

27. What is the value of the land in Ex. 26, at $54.25 per acre? *Ans.* 1672.25625.

28. What part of a mile is 360 yd.? *Ans.* $\frac{9}{44}$ mi.

29. What part of a mile is 8 rd. 3 yd. 1 ft.?
Ans. $\frac{11}{640}$ mi.

30. What will 3250 lb. hay cost at $8 per ton?
$\frac{3250}{2000}$ T. $= \frac{13}{8}$ T. *Ans.* $13.

31. How many tons in 8750 lb.?
Ans. $4\frac{3}{8}$ T. $= 4.375$ T.

32. What is the time between May 9th, 1858, and Sept. 11th, 1872? *Ans.* 14 yr. 4 mo. 2 da.

33. What is the time between July 12th, 1863, and Jan. 1st, 1871? *Ans.* 7 yr. 5 mo. 19 da.

DENOMINATE NUMBERS. 127

34. What is the time between Sept. 15th, 1875, and March 10th, 1878? *Ans.* 2 yr. 5 mo. 25 da.

35. George was born April 12th, 1823, and died June 9th, 1879; what was his age?
Ans. 56 yr. 1 mo. 27 da.

36. James was born August 14th, 1854; how old is he July 1st, 1875? *Ans.* 20 yr. 10 mo. 17 da.

37. How many days from March 15th to June 21st of the same year? *Ans.* 98 days.

38. How many days from Jan. 12th to April 10th of the same year?

Ans. $\begin{cases} 88 \text{ days when not leap year;} \\ 89 \text{ days when leap year.} \end{cases}$

REM.—When the time is long, that is, a year or more, we use 30 days for a month and 12 months for a year; but when the time is short, only a few months, we often require the exact number of days.

39. How much land in 4 lots; the first containing 5 A. 3 R. 20 P., the second 4 A. 2 R. 15 P., the third 7 A. 1 R. 5 P., and the fourth 12 A. 15 P.?
Ans. 29 A. 3 R. 15 P.

40. I have a lot of 15 A. 3 R. and 30 P., that I wish to divide into 4 lots of equal size; how much in each?
Ans. 3 A. 3 R. 37½ P.

41. A man having 1236 A. 3 R. 12 P. of land, sold three parcels; the first, 276 A. 1 R. 8 P.; the second, 301 A. 2 R.; the third, 205 A. 12 P.; and the fourth, 101 A. ½ P.; how much remained unsold? *Ans.* 352 A. 3 R. 31½ P.

42. How many acres in 4 farms, each of 201 A. 2 R. 15 P.? *Ans.* 806 A. 1 R. 20 P.

43. How many grains in 12 lb. 3 ℥ 4 ʒ 2 ∋ 12 gr.?
Ans. 70852 grains.

44. How many grains in 8 lb. 9 oz. 10 pwt. 8 gr.?
Ans. 50648 grains.

45. Divide 8 lb. 9 oz. 10 pwt. 8 gr. into four equal parts; how much in each part?
Ans. 2 lb. 2 oz. 7 pwt. 14 gr.

46. A man owns three tracts of land; the first tract contains 546 A. 3 R. 15 P., the second 612 A. 2 R. 25 P., and the third 408 A. 3 R. 12 P.; he devises 212 A. to his wife, and the balance equally to his four children; how much will each child get? *Ans.* 339 A. 13 P.

47. How many pills of ¼ grain each, can a druggist make out of 1 lb. 9 ℥ 5 ʒ 2 ∋ 15 gr. of morphia?
Ans. 41740 pills.

48. A man has property valued as follows: one house £2145 9s. 8d., another £1576 16s. 4d., a third £654 15s. 10d.; he has a draft for £2176 9s. 6d., and cash £82 5s. 4d., all of which he wishes to divide into four parts as follows: the second to be double the first, the third once and one-half the second, and the fourth once and one-third the second; how much in each part?

Ans. £663 11s. 8d. = 1st part;
£1327 3s. 4d. = 2d part;
£1990 15s. = 3d part;
£2654 6s. 8d. = 4th part.

RATIO.

Two fractions can be formed with any two integral numbers, the one a proper fraction and the other an improper fraction; thus, $\frac{7}{9}$ and $\frac{9}{7}$ can be formed with 7 and 9. When the proper fraction is a multiplier of any number, the product is less than the number multiplied; therefore, this fraction is termed a *Diminishing Ratio*. But when the improper fraction is a multiplier, the product is greater than the multiplicand; hence, the improper fraction is termed an *Increasing Ratio*.

PROBLEMS.

1. If 5 lbs. sugar cost 50 cents, what will 9 lbs. cost?

It is evident that 9 lbs. will cost more than 5, and just as much more as is indicated by the increasing ratio formed by the two like terms, 5 lbs. and 9 lbs.

If 5 lbs. cost 50 cts., 9 lbs. will cost $\cancel{50}$ cts. $\times \frac{9}{5} = 90$ cts.; this may be further demonstrated thus,

$$5 \text{ lbs.} = 50 \text{ cts.}$$
$$1 \text{ lb.} = 10 \text{ cts.}$$
$$9 \text{ lbs.} = 90 \text{ cts.}$$

In a problem of ratios, the one ratio is given, and one of the terms of the other ratio, to get the second term; thus, in the above:

RATIO.

Given,   5 lbs. sugar and 50 cents.
Required, 9 lbs. sugar and   ?

The ratio of the money will be the same as of the sugar. As the required sugar is more than the given, the ratio must be increasing; that is, $\frac{9}{5}$. ∴ $50 \times \frac{9}{5} = 90$, the ratio of the required money to the given, $\frac{90}{50} = \frac{9}{5}$, the same as of the sugar.

EXAMPLES.

1. If 5 bushels of wheat cost $6.25, what will 8 bushels cost?

Given   5 bu. and $6.25.
Required, 8 bu. and

$$\$6.25 \times \tfrac{8}{5} = \$10.00.$$

$$\frac{8}{5} = 25\overline{)\frac{1000}{625}} = 5\overline{)\frac{40}{25}} = \frac{8}{5}.$$

The ratios of the wheat and of the money is the same.

REM.—Ratios can only be formed by two like terms.

2. If 5 bushels of oats cost $1.50, what will 21 bushels cost?

Given   5 bu.   and $1.50.
Required, 21 bu. and

$$\$1.50 \times \tfrac{21}{5} = \$6.30.$$

$\frac{21}{5} = \frac{126}{30} = \frac{21}{5}$; the ratio is the same.

REM.—Write the given terms in a line and the like term of the required immediately under that of the given. One term of the required is wanting, and the given like term may be called the term of demand, and should be placed first and multiplied by the ratio, having for its numerator the required term of the ratio, and for its denominator the given term.

RATIO. 131

As a general thing, an increase in the required term of the ratio will take more of the unknown to accomplish it; an increased amount of goods will cost a greater sum of money; an enlarged piece of work, an additional sum of money; and the greater the work, the longer time to perform it, etc. In examples of this kind, the ratios are direct, and the required term of the ratio holds the place of the numerator and the given term that of the denominator, and the product of the ratio and the odd given term is the term required.

3. If a man travel 40 miles in 8 hours, how many miles will he travel at that rate in 18 hours?

  Given 40 miles. and 8 hours.
  Required, ? miles and 18 hours.

$$\cancel{40} \text{ miles} \times \tfrac{\cancel{18}}{\cancel{8}} = 90 \text{ miles.}$$

4. If 15 bushels of wheat yield 3 barrels of flour, how many bushels will yield 10 barrels of flour?

  Given 15 bu. and 3 barrels of flour.
  Required, ? and 10 barrels of flour.

     SOLUTION.

$$\cancel{15} \times \tfrac{\cancel{10}}{\cancel{3}} = 50 \text{ bushels.}$$

5. If a man travel 30 miles in 2 days, how long will it take him to travel 240 miles?

  Given 30 miles and 2 days.
  Required, 240 miles and ?

6. If a staff 4 feet long cast a shadow 3 feet, what is the height of a steeple which casts a shadow 90 feet?

	STAFF.	SHADOW.
Given	4 ft.	3 ft.
Required,	?	90 ft.

RATIO.

7. If the interest of $100 for one year is $5, what would be the interest of $500 for the same time?

	PRINCIPAL.		INTEREST.
Given	$100	and	$5.
Required,	$500	and	?

8. If $\frac{2}{3}$ of a barrel of flour cost $4, what will $4\frac{2}{3}$ barrels cost?

Given	$\frac{2}{3}$ barrel	and	$4.
Required	$4\frac{2}{3}$ barrel	and	?

$$4 \times \frac{4\frac{2}{3}}{\frac{2}{3}}. \qquad \$4 \times \frac{\overset{7}{\cancel{14}}}{\cancel{3}} \times \frac{\cancel{3}}{\cancel{2}} = \$28.$$

REM. $4\frac{2}{3}$ is a multiplier, and $\frac{2}{3}$ is a divisor; the $\frac{2}{3}$ must be inverted.

9. If $4\frac{1}{2}$ bushels of wheat cost $5.40, what will $8\frac{3}{4}$ bu. cost? $9\frac{1}{4}$ bu.? $23\frac{3}{4}$? $31\frac{2}{3}$? $47\frac{1}{4}$? $39\frac{1}{4}$? $58\frac{3}{4}$? $97\frac{1}{4}$? $106\frac{3}{4}$?

10. If 8 bushels of wheat cost $10, what will be the cost of $9\frac{1}{2}$ bu.? $10\frac{1}{4}$ bu.? $15\frac{2}{3}$ bu.? $37\frac{1}{4}$ bu.? $95\frac{2}{3}$ bu.? $125\frac{3}{4}$ bu.? $150\frac{1}{2}$ bu.? $279\frac{2}{3}$ bu.?

11. If $5\frac{3}{4}$ acres of land cost $230, what is the cost of $6\frac{1}{4}$ acres? $7\frac{3}{4}$ acres? $12\frac{1}{4}$ acres? $13\frac{1}{4}$ acres? $17\frac{2}{3}$? $18\frac{2}{3}$? $19\frac{1}{4}$? $20\frac{2}{3}$? $37\frac{2}{3}$? $49\frac{1}{4}$?

12. If $2\frac{1}{3}$ acres of land cost $110, what will $\frac{1}{2}$ of an acre cost? $\frac{1}{4}$ acre? $\frac{1}{3}$ acre? $\frac{1}{5}$ acre? $\frac{2}{3}$ acre? $\frac{3}{4}$ acre? $\frac{4}{5}$ acre? $1\frac{1}{4}$ acres? $1\frac{2}{3}$ acres?

13. If $\frac{5}{12}$ of a yard of cloth cost $\frac{9}{10}$ of a dollar, what will $\frac{2}{7}$ of a yard cost?

Given	$\frac{5}{12}$ yd.	and	$\$\frac{9}{10}$.
Required,	$\frac{2}{7}$ yd.	and	?

$$\frac{9}{10} \times \frac{2}{7} \times \frac{12}{5}.$$

RATIO.

14. If ½ yard of cloth cost $2, what will 3 ells F. cost?

$$2 \times \tfrac{9}{8} =$$

15. If ¾ yard of cloth cost $2.25, what will 5 ells English cost? What will 5 ells French cost?

$$2.25 \times \tfrac{25}{3}.$$

PRACTICAL EXAMPLES.

1. If 12 bushels of wheat cost $15, what will 42 bushels cost?

  Given 12 bushels and $15.
  Required 42 bushels and ?

$$15 \times \tfrac{42}{12} = 15 \times \tfrac{7}{2} = \tfrac{105}{2} = \$52\tfrac{1}{2}.$$

2. If 15 bushels of oats cost $4.50, what will 75 bushels cost?

  Given 15 bushels and $4.50.
  Required 75 bushels and ?

$$\$4.50 \times \tfrac{75}{15}.$$

3. If 12 bushels of wheat cost $15, how many bushels will $52½ buy?

  Given 12 bushels and $15.
  Required ? and $52.50.

$$12 \times \frac{52\tfrac{1}{2}}{15} = \cancel{12} \times \tfrac{105}{\cancel{15}} = 42 \text{ bushels.}$$

4. If 15 bushels of oats cost $4.50, how many bushels will $22.50 buy?

  Given 15 bushels and $4.50.
  Required ? and $22.50.

$$15 \times \tfrac{22.50}{4.50} = 75 \text{ bushels.}$$

5. If a man travel 48 miles in 6 hours, how many miles will he travel in 56 hours at the same rate?

Ans. 448 miles.

6. If 20 bushels of wheat yield 4 barrels of flour, how many bushels will be required for 15 barrels of flour?

Ans. 75 barrels.

7. If 20 bushels of wheat yield 4 barrels of flour, how many barrels of flour will 75 bushels yield?

8. If a man travel 50 miles in 2 days, how long will it take him to perform a journey of 375 miles at the same rate?

9. If a staff 3 feet long cast a shadow 4 feet, what will be the length of the shadow of a steeple 180 ft. high?

Given   3 ft., staff   and   4 ft., shadow.
Required 180 ft., steeple and   ?   shadow.

Shadow of steeple 240 ft.

10. If the interest of $100 for one year is $6, what is the interest of $500 for the same time?   *Ans.* $30.

11. If a man's salary amounts to $1500 in 3 years, how much will it amount to in 7 years?

12. If $12\frac{1}{2}$ bushels of wheat cost $15, what will $35\frac{1}{2}$ bushels cost?

$$\frac{35\frac{1}{2} \times 2}{12\frac{1}{2} \times 2} = \frac{71}{25}. \qquad 15 \times \frac{71}{25}.$$

13. If $\frac{2}{3}$ of a barrel of flour cost $6, what will $5\frac{1}{4}$ barrels cost?

The $5\frac{1}{4}$ is a multiplier and the $\frac{2}{3}$ is a divisor.

$$6 \times \frac{3}{2} \times \frac{21}{4}.$$

RATIO.

14. If $5\frac{1}{4}$ bushels of wheat cost $6.60, what will $37\frac{1}{3}$ bushels cost?

$$\$.60 \times \overset{\cdot 20}{\tfrac{2}{35}} \times 1\tfrac{12}{4} = \$44.80.$$

15. If $\frac{4}{5}$ of an acre of land is worth $36.40, what is the value of $15\frac{3}{16}$ acres at the same rate?

$$15\tfrac{3}{16} = \cdot\tfrac{243}{16}. \qquad \$36.40 \times \tfrac{5}{4} \times \tfrac{243}{16}.$$

16. If $7\frac{4}{5}$ acres of land cost $1560, what will $\frac{3}{4}$ of an acre cost? *Ans.* $150.

17. The assets of a firm are $48576 and the liabilities $97648; how much will a creditor receive to whom is owing $1654?

$$48576 \times \tfrac{1654}{97648} =$$

18. If $\frac{7}{12}$ of a yard cost $\$\frac{7}{20}$, what will $\frac{4}{5}$ of a yard cost? *Ans.* $\$\frac{12}{5}$.

19. If $\frac{7}{14}$ of a ship cost £49, what is $\frac{4}{7}$ of it worth? *Ans.* £32.

20. What will 57 acres 3 rods and 24 perches of land cost, if 5 acres 2 rods and 6 perches cast $224.64?
Reduce to perches. *Ans.* $2348.83+.

21. If $1.50 will buy 6 lbs. sugar, how much sugar will $36.50 buy?

22. If $5\frac{1}{4}$ yards of cloth cost $7.15, what will $72\frac{1}{2}$ yards cost? $7.15 × $\tfrac{145}{11}$.

23. If $4\frac{1}{4}$ lbs. ham cost 87 cents, what will $35\frac{1}{4}$ lbs. cost? *Ans.* $6.79+.

24. If .75 of a ton of hay cost $15, what will 2.25 tons cost? *Ans.* $45.

25. If 100 acres of land cost $400, what will 830 acres cost?

26. If $\frac{5}{11}$ of a ship cost $4500, what will $\tfrac{11}{20}$ cost?

27. If .25 of an acre of land is worth $32, what is .875 of an acre worth?

28. If 5¾ yards of muslin cost 46 cents, what will 20¼ yards cost? *Ans.* 162 cents.

29. If 3 yards of cloth cost $15.75, what will 6 ells English cost? *Ans.* $39.37½.

30. If 5 ells Fl. cost $15.15, what will 6 ells Fr. cost?

In the preceding examples, the ratios were all direct; as in those cases any increase in the required term of the ratio demanded a similar increase of the unknown; but there are cases which require the ratio to be inverted, such as, the more men employed, the less time will be required to perform a piece of work; the more hours employed in the day, the less days; the wider the material, the less yards it will take to make a garment, etc.

These cases of inverse ratio are readily detected by asking this question: "Will an increase of the required term of the ratio demand an increase in the unknown term?" If it does, the ratio is direct; but if an increase in the required term of the ratio demand a diminution of the unknown term, the ratio must be inverted; thus,

PROBLEM.

If 4 men can do a piece of work in 10 days, how long will it take 8 men to do the same work?

$$\text{Given} \quad \text{4 men in 10 days.}$$
$$\text{Required,} \quad \text{8 men in} \quad ?$$

$$\cancel{10}^{5} \text{ days} \times \tfrac{\cancel{4}}{\cancel{8}_2} = 5 \text{ days.}$$

RATIO.

COR.—It is evident that 8 men will do it in less time; that is, in one-half the time that it will take 4, which ratio is expressed by the diminishing ratio of 4 and 8, that is, $\frac{4}{8} = \frac{1}{2}$, in which the given term is the numerator of the ratio and the required term the denominator.

EXAMPLES.

1. If 5 men can dig a ditch in 20 days, how many men will dig it in 25 days?

$$\text{Given} \quad 5 \text{ men} \quad \text{and} \quad 20 \text{ days.}$$
$$\text{Required,} \quad ? \quad \text{and} \quad 25 \text{ days.}$$

$$\cancel{5} \text{ men} \times \frac{\cancel{20}^{4}}{\cancel{25}_{5}} = 4 \text{ men.}$$

REM.—An increase in the 25 will require less men.

2. If 6 horses eat a certain quantity of hay in 30 weeks, how many horses will consume the same quantity of hay in 9 weeks?

$$\text{Given} \quad 6 \text{ horses} \quad \text{and} \quad 30 \text{ weeks.}$$
$$\text{Required,} \quad ? \quad \text{and} \quad 9 \text{ weeks.}$$

$$\cancel{6}^{2} \text{ horses} \times \frac{\cancel{30}^{10}}{\cancel{9}_{3}} = 20 \text{ horses.}$$

3. If a man perform a journey in 12 days, when the days are 9 hours long, how many days of 12 hours will it take him?

$$\text{Given} \quad 12 \text{ days} \quad \text{and} \quad 9 \text{ hours.}$$
$$\text{Required,} \quad ? \quad \text{and} \quad 12 \text{ hours.}$$

4. If 10 men can build a wall in 40 days, how many men will be required to build the same wall in 10 days?

$$\text{Given} \quad 10 \text{ men and } 40 \text{ days.}$$
$$\text{Required} \quad ? \quad \text{and } 10 \text{ days.}$$

$$\cancel{10} \times \cancel{\tfrac{40}{10}} = 40 \text{ men.}$$

An increase in the required term of the ratio demands a diminution in the unknown.

5. If 12 horses eat a certain quantity of hay in 54 weeks, how many horses will consume the same hay in 9 weeks? *Ans.* 72 horses.

6. If a man perform a journey in 24 days when the days are 9 hours long, how many days will it take him when the days are 12 hours long? *Ans.* 18 days.

7. If 10 men reap 30 acres of wheat in 3 days, how long will it take 5 men to reap the same field? *Ans.* 6 days.

COMPOUND RATIO.

When there are two or more ratios, it is termed *Compound Ratio;* thus,

If 3 men in 12 days build 40 rods of wall, how many rods will 9 men build in 24 days?

$$\text{Given} \quad 3 \text{ men, } 12 \text{ days, } 40 \text{ rods.}$$
$$\qquad \quad 9 \text{ men, } 24 \text{ days, } \quad ?$$

$$40 \times \tfrac{\overset{3}{9}}{\underset{1}{3}} \times \tfrac{\overset{2}{24}}{12} = 240 \text{ rods.}$$

Rem.—Each ratio is direct.

If 12 men dig a ditch 20 rods long in 18 days by working 8 hours a day, how many men will dig a ditch 40 rods long in 24 days, working 6 hours a day?

Given   12 men,   20 rods,   18 days,   8 hours.
            ?   40 rods,   24 days,   6 hours.

$$12 \text{ men} \times \frac{40}{20} \times \frac{18}{24} \times \frac{8}{6} = 24 \text{ men.}$$

EXEMPLIFICATION.—The longer the trench, the more men it will take, and the ratio is direct; but the greater the number of days and the more hours of each day, the fewer men would be required; hence these two ratios are inverse.

COR.—Each ratio must be dealt with as in the preceding article.

EXAMPLES.

1. If 4 men in 12 days build 40 rods of wall, how many rods will 6 men build in 18 days?

Given   4 men,   12 days,   40 rods.
Required   6 men,   18 days,   ?

$$40 \text{ rods} \times \frac{6}{4} \times \frac{18}{12} = 90 \text{ rods.}$$

The ratios are all direct.

2. If 18 men dig a trench 30 rods long in 24 days by working 8 hours a day, how many men will dig a trench 60 rods long in 64 days working 6 hours a day?

Given   18 men,   30 rods,   24 days,   8 hours.
Required   ?   60 rods,   64 days,   6 hours.

$$18 \times \frac{60}{30} \times \frac{24}{64} \times \frac{8}{6} = 18 \text{ men.}$$

Observe, the longer the trench, the more men will be

required; but the more days the less men, and the more hours the less men; the first ratio is direct, the other two inverse.

3. If 6 men in 16 days of 9 hours each, build a wall 20 feet long 6 feet high and 4 feet thick, in how many days of 8 hours each will 24 men build a wall 200 ft. long 8 ft. high and 6 ft. thick?   Ans. 90 days.

4. If 12 men mow 24 acres of grass in 2 days of 10 hours each, how many hours a day must 16 men work to mow 80 acres in 4 days?

Given   12 men, 24 acres, 2 days, 10 hours.
Required 16 men, 80 acres, 4 days,   ?

$$\overset{5}{\cancel{10}} \times \tfrac{12}{16} \times \tfrac{80}{24} \times \underset{2}{\tfrac{2}{4}} = \tfrac{25}{2} = 12\tfrac{1}{2} \text{ hours.}$$

5. If $100 in 12 months gain $6, how long will it take $600 to gain $24?

Given $100 prin., 12 mo., $6 interest.
Required $600 prin.,   ? $24 interest.

$$\overset{2}{\cancel{12}} \times \tfrac{1}{6} \times \tfrac{4}{1} = 8 \text{ months.}$$

6. If 8 horses eat 42 bushels of oats in 24 days, how many bushels will suffice 16 horses 36 days?
                                                     Ans. 126 bushels.

7. If it cost $30 to transport 6 cwt. 2 qrs. 180 miles, what will be the cost of transportation of 19 cwt. 2 qrs. 270 miles?   Ans. $135.

8. If 42 men in 270 days of 8¼ hours each can build a wall 98¾ yards long 7½ ft. high and 2⅓ ft. thick, in how many days of 11⅓ hours each can 63 men build a wall 45¼ yds. long 6$\tfrac{7}{12}$ ft. high and 3¼ ft. thick?

$270 \times \tfrac{42}{63} \times \tfrac{17}{2} \times \tfrac{3}{34} \times 1\tfrac{10}{3} \times \tfrac{12}{15} \times \tfrac{25}{8} \times \tfrac{1}{315} \times \tfrac{2}{15} \times \tfrac{2}{3} = 68 \text{ da.}$

RATIO.   141

REM.—When all the terms are arranged in ratios, the cancellation is more easily performed. It is better, however, first to arrange the like terms together, and then mark the direct and indirect ratios.

9. If 14 men can reap 84 acres in 6 days, how many men must be employed to reap 44 acres in 4 days?

10. A wall 600 feet in length is to be built in 30 days; 10 men have been employed at it for 12 days, and have built 240 feet. How many more men must be employed in order to finish it in the given time?

Ans. The 10 men will finish it in the given time.

11. If 12 men make 600 pairs of shoes in 30 days, how many men will make 12000 pairs in 90 days?

Ans. 80 men.

SIMPLE EQUATIONS.

An *Equation* consists of two equal members placed opposite each other with the sign of equality between them. The members are called the right and left hand members. Thus, $8 + 4 = 6 + 6$.

The analysis of simple ratios may be rendered by equations, thus:

1. If 12 bushels of wheat cost $15, what will 42 bushels cost?

	12 bu. = $15.
Divide both members by 12.	1 bu. = $1.25.
Multiply both members by 42.	42 bu. = $52.50.

That is, if 12 bushels equals or costs $15, one bushel will cost $\frac{1}{12}$ of $15; that is, $1.25 and 42 bushels will cost $1.25 × 42 = $52.50.

RATIO.

Ax. 1. If equals be multiplied by equals, the products will be equal.

Ax. 2. If equals be divided by equals, the quotients will be equal.

2. If 15 bushels of oats cost $4.50, what will 75 bushels cost?

$$15 \text{ bu.} = \$4.50.$$
Divide both members by 15. $1 \text{ bu.} = .30.$
Multiply both members by 75. $75 \text{ bu.} = \$22.50.$

3. If 20 bushels of wheat yield 4 barrels of flour, how many bushels will yield 15 barrels of flour?

$$4 \text{ barrels} = 20 \text{ bushels.}$$
$$1 \text{ barrel} = 5 \text{ bushels.}$$
$$15 \text{ barrels} = 75 \text{ bushels.}$$

4. A man bequeathed his estate of $10000 to his son and daughter; the son to have $2000 more than the daughter. What was the share of each?

The work may be shortened by letting x represent one of the unknowns; thus,

Let x = daughter's share.
             $x + 2000$ = son's share.
Add the 2 shares, $2x + 2000 = 10000$
                $2x = 8000$
                 $x = 4000$ daughter's share.
       $x + 2000 = 6000$ son's share.

Cor. 1. If equals be subtracted from equals, the remainders will be equal.

2. If equals be divided by equals, the quotients will be equal.

5. A and B hired a pasture for $55; A paid 13 dollars more than B. What did each pay?

RATIO. 143

Let x = what B paid.
Then $x + 13$ = what A paid.
$2x + 13 = 55$
$2x = 42$
$x = 21$ = what B paid.
$x + 13 = 34$ = what A paid.

EQUATION OF PAYMENTS

consists in averaging the time of several payments so as to get the time when all may be paid at once, without loss to either party; thus,

1. Bought goods for $300 payable in 2 months, $500 in 3 months, $700 in 4 months. At what time should the whole be paid without loss to either party? $\frac{49}{15} = 3\frac{4}{15}$

300×2 = $600 in 1 month.
500×3 = $1500 in 1 month.
700×4 = $2800 in 1 month.
$1500 \times 3\frac{4}{15}$ = $4900 in 1 month.

The discount of 4900 for 1 month = 1500 in $3\frac{4}{15}$ months. The whole should be paid in 3 mo. 8 da., *Ans.*

2. Bought goods for $500 payable in 30 days, $600 in 60 days, $1000 in 90 days, $5000 in 120 days. What would be the equated time to pay the whole?

500×30 = 15000
600×60 = 36000
1000×90 = 90000
5000×120 = 600000
$71|00 \times 104\frac{22}{71}$ = 71) $7410|00$ ($104\frac{22}{71}$ days.
 71

 310
 284

 22

AVERAGING ACCOUNTS.

When sales are made at different times and on different terms, to find on mean time when all may be paid without loss to either party; thus,

1. A merchant sells goods as follows:

       Jan.   1st, $100 on 1 month.
       Jan. 18th, 200 on 1 month.
       Feb.   1st, 300 on 2 months.
       Feb. 12th, 250 on 3 months.

At this date no payment has been made. In how many days should the whole be paid at once in order to secure both parties?

Begin at Feb. 1st, when the first account is due.

```
   Feb.  1st,  $100 ×   0
   Feb. 18th,   200 ×  17 =  3400
   Apr.  1st,   300 ×  59 = 17700
   May  12th,   250 × 100 = 25000
                ───                
                850        ) 46100 ( 54 days.
                             4250
                             ─────
                             3600
                             3400
                             ─────
                              20|0    4
                              ──── = ──
                              85|0   17
```

Due 54 days after Feb. 1st; that is, 27th March.

2. When an account has both debits and credits, begin with the first date.

RATIO.   145

Dr. B. Thompson in acccount with I. Parsons.   Cr.
To invoice of goods due
 March 15th, $400 March 1st, by cash, $300
 April 1st, 300 April 1st, by cash, 200
 April 15th, 200 April 15th, by cash, 300
 May 1st, 400 May 15th, by cash, 500
 May 15th, 600
Before conjoined.

$400 × 14 = $5600   300 × 0
 300 × 31 = 9300   200 × 31 = $6200
 200 × 45 = 9000   300 × 45 = 13500
 400 × 61 = 24400   500 × 75 = 37500
 600 × 75 = 45000 $1300 $57200
$1900 $93300
 1300   57200
 $600 6) 361|00
     60 days.

CONJOINED EQUATIONS AND RATIOS

consist of a number of equations, and in each successive equation the left hand member or antecedent is a like term of the right hand member or consequent of the preceding equation.

PROBLEMS.

1. If 5 oranges are worth 8 lemons, 3 lemons worth 10 apples, 4 apples worth 1 melon, and 6 melons worth 75 cents, how much are 12 oranges worth?

Arrange in equations.

       oranges.
How many cents = 12 lemons.
     5 = 8 apples.
      3 = 10 melon.
       4 = 1 cents.
        6 = 75.

$$
\begin{aligned}
6 \text{ melons} &= 75 \\
1 \text{ melon} &= \tfrac{75}{6} \\
4 \text{ apples} &= \tfrac{75}{6} \\
1 \text{ apple} &= \frac{75}{6 \times 4} \\
10 \text{ apples} &= 75 \times \tfrac{1}{6} \times \tfrac{10}{4} = 3 \text{ lemons.} \\
8 \text{ lemons} &= 75 \times \tfrac{1}{6} \times \tfrac{10}{4} \times \tfrac{8}{3} = 5 \text{ oranges.} \\
12 \text{ oranges} &= 75 \times \tfrac{1}{6} \times \tfrac{10}{4} \times \tfrac{8}{3} \times \tfrac{12}{5} = 25 \times 2 \times 4 = 200
\end{aligned}
$$

2. If 6 cords of wood buy 12 barrels of apples, 8 barrels of apples buy 6 barrels of oranges, 2 barrels of oranges 32 lbs. butter, 40 lbs. butter = 1 ton coal, and 6 tons coal = 15 barrels flour, how many cords wood will 12 barrels flour buy?

wood. apples.
$6 = 12$ oranges.
$\quad 8 = 6$ butter.
$\quad\quad 2 = 32$ coal.
$\quad\quad\quad 40 = 1$ flour.
$\quad\quad\quad\quad 6 = 15$
$\quad\quad\quad\quad\quad 12 =$ how many cd. wood?

$$
\begin{aligned}
1 \text{ bbl. apples} &= \tfrac{6}{12}. \\
8 \text{ bbls. apples} &= 6 \times \tfrac{8}{12} = 6 \text{ bbls. oranges.} \\
2 \text{ bbls. oranges} &= 6 \times \tfrac{8}{12} \times \tfrac{2}{2} = 32 \text{ lbs. butter.} \\
40 \text{ lbs. butter} &= 6 \times \tfrac{8}{12} \times \tfrac{2}{2} \times \tfrac{40}{32} = 1 \text{ ton coal.} \\
6 \text{ tons coal} &= 6 \times \tfrac{8}{12} \times \tfrac{2}{2} \times \tfrac{40}{32} \times \tfrac{6}{1} = 15 \text{ bbls. flour.} \\
12 \text{ bbls. flour} &= 6 \times \tfrac{8}{12} \times \tfrac{2}{2} \times \tfrac{40}{32} \times \tfrac{6}{1} \times \tfrac{12}{15} \\
&= 2 \times 2 \times 2 = 8 \text{ cords of wood.}
\end{aligned}
$$

Cor. 1.—When the odd term, which is the term of demand, is the consequent of the last equation, make it

RATIO. 147

the first term and multiply it by all the other terms as ratios in the order in which they stand, the upper like member as the numerator and the lower one as denominator.

COR. 2.—When the antecedent of the first equation is the odd term, multiply it by all the other terms inverted as ratios.

EXAMPLES.

1. If A can do as much work in 3 days as B in $4\frac{1}{2}$ days, B as much in 9 days as C in 12 days, and C as much in 10 days as D in 8 days, how many days work of D's is equal to 5 days of A's.

$$\begin{array}{rl} \text{What D} = & \begin{array}{c} A \\ 5 \end{array} \\ & 3 = \begin{array}{c} B \\ 4\frac{1}{2} \end{array} \\ & \qquad 9 = \begin{array}{c} C \\ 12 \end{array} \\ & \qquad\qquad 10 = \begin{array}{c} D \\ 8. \end{array} \end{array}$$

$$8 \times \frac{\cancel{12}}{\cancel{10}} \times \frac{4\frac{1}{2}}{\cancel{9}} \times \frac{\cancel{5}}{\cancel{3}} = 8.$$

Ans. 5 days of A. = 8 days of D.

2. If 4 men can do as much work as 5 women, 6 women as much as 9 boys, 15 boys as much as 25 girls, and 27 girls can bind 300 sheaves in an hour, how many sheaves can 18 men bind in the same time?

$$\begin{array}{rl} & \text{men.} \\ ?\text{ sheaves} = & 18 \quad \text{women.} \\ & \quad 4 = 5 \quad \text{boys.} \\ & \qquad 6 = 9 \quad \text{girls.} \\ & \qquad\quad 15 = 25 \quad \text{sheaves.} \\ & \qquad\qquad 27 = 300 \end{array}$$

$$300 \times \tfrac{18}{4} \times \tfrac{9}{15} \times \tfrac{5}{6} \times \tfrac{25}{27} =$$

148   RATIO.

3. If 1 pound is worth 20 shillings, 1 shilling = 12 pence, and 1 penny = 4 farthings, how many farthings are there in 1 pound?

farthings.   pence.
4   =   1   shilling.
          12 = 1   pound.
                   20 = 1
                            1 = ? farthings.

$4 \times \frac{12}{1} \times \frac{20}{1}$ = 960 farthings = 1 pound.

4. If 3 lbs. tea are worth 7 lbs. coffee, 14 lbs. coffee worth 48 lbs. sugar, and 18 lbs. sugar worth 27 lbs. soap, how many lbs. soap are 6 lbs. tea worth?

               tea.
? soap =   6   coffee.
              3 =   7   sugar.
                      14 = 48
Ans. 72 lbs. soap.   18 = 27 soap.

5. If 1 French crown = 80 pence Holland, 83 pence Holland = 48 pence English, 4 pence English = 70 pence Hamburg, 64 pence Hamburg = 1 florin Frankfort, how many florins Frankfort = 166 French crowns?
                                       Ans. 2100.

6. If 30 acres of land in Frederick are worth 40 acres in Washington, 60 acres in Washington are worth 90 acres in Allegheny, 100 acres in Allegheny worth 40 acres in Carroll, 50 acres in Carroll worth 75 in Montgomery, how many acres in Frederick are worth 450 in Montgomery?

acres.   acres.
30 F. = 40 W.   acres.
            60 W. = 90 A.   acres.
                         100 A. = 40 C.   acres.
                                      50 C. = 75 M.
Ans. 375 acres.   450 M. = ? F.

PROPORTION.

Rem. 1.—A proportion may be formed of two or more equal ratios; thus, $\frac{2}{3} = \frac{4}{6}$, ∴ 2 : 3 :: 4 : 6; that is, as 2 is to 3, so is 4 to 6, and as $\frac{2}{3} = \frac{4}{6}$, by multiplying each numerator by the other denominator, the equation becomes 12 = 12; that is, the product of the first and last, called the extremes, is equal to the product of the second and third, called the means; hence, any three of these terms being given, the fourth is readily found.

Rem. 2.—The left-hand member of the equation is the product of the means, and the right member of the extremes.

PROBLEMS.

1. Given $\frac{3}{?} = \frac{6}{4}$ to find the denominator of the first fraction.

Since the first numerator is one-half the second, the first denominator must be one-half the second.

∴ $\frac{3}{2} = \frac{6}{4}$ or 6 : 4 :: 3 : 2, and $\frac{4 \times 3}{6} = 2$, *Ans.*

2. Given $\frac{3}{2} = \frac{6}{?}$, hence 3 : 2 :: 6 : ?

∴ $\frac{2 \times 6}{3} = 4$, *Ans.*

3. Given $\frac{?}{2} = \frac{6}{4}$, hence 4 : 6 :: 2 : ?

∴ $\frac{6 \times 2}{4} = 3$, *Ans.*

Cor.—Arrange the proportion so that the unknown is to be the fourth term; and as the product of the two means is equal to the product of the two extremes, therefore the unknown is equal to the quotient obtained by dividing the product of the two means by the given extreme.

4. If 5 lb. of sugar cost 50 cents, what will 25 lb. cost?

As 5 : 25 :: 50 : ? $\dfrac{\cancel{25}^{5} \times 50}{\cancel{5}} = 250$ cts., *Ans.*

REM.—It is evident that the ratio of the money must be the same as that of the sugar.

5. If 4 hats cost $12, what will 15 hats cost?

As 4 : 15 :: 12 : ? $\dfrac{15 \times \cancel{12}^{3}}{\cancel{4}} = \$45$, *Ans.*

6. If a man walk 75 miles in 3 days, how many miles will he walk in 21 days?

3 : 21 :: 75 : ? ∴ $\dfrac{\cancel{21}^{7} \times 75}{\cancel{3}} = 525$ miles, *Ans.*

7. If 4 gallons of syrup cost $2, what will 64 gallons cost?

4 : 64 :: $2 : $? ∴ $\dfrac{\cancel{64}^{16} \times 2}{\cancel{4}} = \$32$, *Ans.*

REM. 1.—In a proportion each ratio is composed of two like quantities termed a couplet; the couplets, however, are generally unlike quantities. If the like quantities are of different denominations, they must be reduced to the same denomination.

REM. 2.—In any number of proportions the odd terms are called antecedents and the even terms consequents.

REM. 3.—The proportion is correct, although the ratios are inverted by alternating the antecedents and consequents.

EXAMPLES.

1. If a railroad car runs 21 miles in 50 minutes, how far will it run in 5 hours and 50 minutes?

As 50 m. : 5 h. 50 m. :: 21 miles : ? miles.

$50 : 350 :: 21 : ?$ $\therefore$ $\dfrac{350 \times 21}{50} = 147$ miles, *Ans.*

2. If 45 acres of land can be purchased for $900, what is the cost of 175 acres at the same rate? *Ans.* $3500.

3. If a man can do a piece of work in 24 days, working 10 hours a day, how long will it take him to do the same, if he works 12 hours a day? *Ans.* 20 days.

4. If the wages of 5 men for 20 days is $125, what would be the wages of 8 men for 24 days? *Ans.* $240.

Rem.—5 men for 20 days is the same as 100 men for one day, and 8 men for 24 days is the same as 192 men for one day.

5. If a man travel 117 miles in 15 days of 9 hours each, how far will he travel in 20 days of 12 hours each?
Ans. 208 miles.

6. If 25 lb. butter purchase 40 lb. cheese, how many pounds butter will purchase 120 lb. cheese?
Ans. 75 lb. butter.

7. Two men engage in business; A puts in $7500 and B $3000; the profits are in proportion to the stock except that B is to receive $500 for his special attention to the business—the profits are $2500; what does each receive?   *Ans.* A receives 1428\frac{4}{7}$ and B 1071\frac{3}{7}$.

8. If $200 gain $12 in one year, what will $600 gain in 9 months? *Ans.* $27.

Rem.—Examples like this should be solved by proportion, by ratio, and by analysis.

PERCENTAGE.

Per Cent. means per hundred, and is generally expressed fractionally; thus, 5 per cent., 6 per cent., marked 5% and 6%, is expressed $\frac{5}{100}$, $\frac{6}{100}$, etc., or .05, .06; thus, $\frac{5}{100}$ of 100 = $\cancel{100} \times \frac{5}{\cancel{100}}$ = 5, and $\frac{6}{100}$ of 100 is 6. $\frac{100}{100}$ = 100%, $\frac{1}{2}$ = 50%, $\frac{1}{3}$ = 33⅓%, $\frac{1}{4}$ = 25%, $\frac{1}{5}$ = 20%, $\frac{1}{6}$ = 16⅔%, $\frac{1}{50}$ = 2%.

EXAMPLES.

1. What is 5% of 200? $\cancel{200} \times \frac{5}{\cancel{100}}$ = 10, *Ans.*
   What is 5% of 300? *Ans.* 15.
   What is 5% of 400? *Ans.* 20.
2. What is 5% of 245? $2.45 \times \frac{5}{\cancel{100}}$ = 12.25, *Ans.*

The 100 is canceled in the 245 by pointing off two places of decimals.

3. What is 6% of 300? $\cancel{300} \times \frac{6}{\cancel{100}}$ = 18, *Ans.*
   What is 6% of 400? *Ans.* 24.
   What is 6% of 500? *Ans.* 30.
4. What is 6% of 368? $3.68 \times \frac{6}{\cancel{100}}$ = 22.08.

COMMISSION, OR BROKERAGE.

The business of a commission merchant or broker is to make purchases and sales, on which he receives a percentage.

PROBLEM I.

A purchase of $100 worth of goods, at 1% commission, will cost $101; that is, $\frac{101}{100}$ of the amount of the purchase.

PERCENTAGE. 153

PROBLEM II.

In a sale of goods for $100, at 1% commission, the owner will realize $99; that is, $\frac{99}{100}$ of the amount of sale.

PROBLEM III.

When stocks, bonds, drafts, or currency, are purchased at a discount of 2%, the cost of $100 worth will be $98; that is, $\frac{98}{100}$ of the face of the bond, etc.; but when they are purchased at a premium of 2%, the cost of $100 worth is $102; that is, $\frac{102}{100}$ of the face.

PROBLEM IV.

In the exchange of currency, when there is a premium on the funds on hand, as that of English money to be exchanged into United States, the premium in favor of England is about 9%; it is computed as follows:

$$\text{Eng. } £ \times \tfrac{40}{9} \times \tfrac{109}{100} = \$ \text{ U. S.,}$$

and $\quad\$ \text{ U. S.} \times \tfrac{9}{40} \times \tfrac{100}{109} = £ \text{ Eng.}$

that is, England gets $109 for every $100 of her money, and the United States must pay $109 of her money for $100 English money.

REM.—This is according to the old exchange value. Now, however, the exchange value of £1 is fixed at $4.86.

EXAMPLES.

1. A broker sold goods to the amount of $6000, at 2% commission, and invested the balance of the proceeds, after deducting 2% on the amount of purchase; what was the owner's portion of the sale, and what amount of goods were purchased?

154   PERCENTAGE.

$6000 \times \tfrac{98}{100} = \$5880 =$ owner's portion of the sale.
$5880 \times \tfrac{100}{102} = \$5764.70\tfrac{10}{102} =$ amt. of goods purchased.

$$\therefore 6000 \times \overset{49}{\underset{\underset{51}{\cancel{100}}}{\cancel{98}}} \times \tfrac{100}{\cancel{102}} = 6000 \times \tfrac{49}{51}$$

$= \$5764\tfrac{16}{51}$, amt. of goods purchased.

Rem.—Observe the difference in the ratios of the sale and purchase.

2. What is the cost of a bond for $5000 at 5% discount, stocks whose face indicate $2000 at 4% premium, $1000 currency at 2% discount, and $3000 gold at 8% premium?

$$\begin{aligned}
\$50\cancel{00} \times \tfrac{95}{\cancel{100}} &= \$4750 \\
20\cancel{00} \times \tfrac{104}{\cancel{100}} &= 2080 \\
10\cancel{00} \times \tfrac{98}{\cancel{100}} &= 980 \\
30\cancel{00} \times \tfrac{108}{\cancel{100}} &= \underline{3240} \\
& \quad \$11050
\end{aligned}$$

Cor.—When brokerage is paid in the exchange of money, the percentage is on the amt. purchased, which, if the rate is 2%, is $\tfrac{100}{102}$ of the funds on hand.

3. If a broker makes sales to the amount of $500, on which he receives 3%, what is his commission?

$$\$\overset{5}{\cancel{500}} \times \tfrac{3}{\cancel{100}} = \$15.$$

2. What is the cost of a draft for $1000 at a premium of ½%?

$$\$\overset{5}{\cancel{1000}} \times \tfrac{100\tfrac{1}{2}}{\cancel{100}} \times \tfrac{201}{\cancel{200}} = \$1005.$$

3. What is the face of a draft at ½% premium, costing $1005?

$$\$\overset{5}{\cancel{1005}} \times \tfrac{100}{\cancel{100\tfrac{1}{2}}} \times \tfrac{200}{\cancel{201}} = \$1000.$$

PERCENTAGE. 155

4. A broker makes sales for $4325, at 2%; what is the brokerage, and what does the owner realize?

$43.25 \times \frac{2}{100} = \$86.50$, commission.

$43.25 \times \frac{98}{100} = \$4238.50$, owner realized.

5. A merchant sells to a broker $3275 uncurrent funds at 5% discount; what does he realize?

163.75×19

$\$327.5 \times \frac{95}{100} = \$3111.25$.

6. An architect charges 1½% for plans and specifications, and 2½% for superintending a building, the cost of which is $10000; what is the architect's fees? *Ans.* $400.

7. A broker has 2% commission, and 3% for guaranteeing payment; what does he receive on sales amounting to $42325? *Ans.* $2116.25.

8. I sent my broker $4000 to purchase goods; what amount of goods did he purchase after deducting commissions at 2% on the amount of goods? $4000 × $\frac{100}{102}$.

For every $102 he gets $100 worth of goods.

9. Bought a draft on New York, the face of it $500 premium, ¼%; what is the whole cost and the premium?

$\$500 \times \dfrac{100\frac{1}{4}}{100} = \dfrac{401}{400} = \quad \$501.25 = $ cost.

$\underline{500.00}$

Premium, $1.25

10. Sold goods to the amount of $4444, and invested the proceeds, after retaining my commissions, which were 2% on the sales, and 1% on the investment; what was the amt. of investment?

$\$4444 \times \frac{98}{100} \times \frac{100}{101} = \$4312$.

PERCENTAGE.

11. A man purchased of a broker 1000 dollars in gold, on which he pays a premium of 15%. How much currency does he invest?

$$\$1000 \times \tfrac{115}{100} = \$1150 \text{ in currency.}$$

12. A merchant sells to a broker $5275 uncurrent funds at 5% discount. What does he realize?

$$\$5275 \times \tfrac{1}{100} \times \tfrac{1}{10} = \quad\begin{array}{l}\$5275.00\\ \underline{263.75}\quad \text{Discount.}\\ \$5011.25\quad \text{Realizes.}\end{array}$$

13. My agent made a compromise with a debtor owing $3264, at 60%, and his fee was 5%. What did I receive?

$$5\% \text{ on } 60 = 3\%; \quad \therefore \quad 3264 \times \tfrac{57}{100} =$$

14. Paid $116.40 commission on the sales of goods, at 1½%. What was the amount of sales? and also the net proceeds?

$$116.40 \times \frac{100}{1\frac{1}{2}} = 116.40 \times \tfrac{200}{3} = 38.80 \times 200.$$

15. A tax collector receives $2430 for his collections, at 2¼%. What is the amount of taxes collected?

$$\$2430 \times \frac{100}{2\frac{1}{4}} = \tfrac{400}{9} \times 2430 = 270 \times 400.$$

16. I sent my broker $5000 to purchase goods, his charge was 2% on the amount of goods. What amount did he buy?

For $102 he bought $100 goods; $\therefore$ $5000 \times \tfrac{100}{102} =$

17. Goods sold on commission at 2¼%, the consignee received $3755.40. What was the amount of sales?

ANALYSIS. $\frac{97\frac{1}{2}}{100} = \frac{195}{200} = \frac{39}{40} = 3755.40.$

$\frac{1}{40} = \frac{3755.40}{39}.$

$\frac{40}{40} = 1 = 3755.40 \times \frac{40}{39}.$

The ratio is $3755.40 \times \frac{100}{97\frac{1}{2}} = \frac{200}{195} = \frac{40}{39}.$

Cancel and compute.

18. A broker sells 50 tons of hay, proceeds $862.944, brokerage 1¼%; how much per ton did the owner realize? *Ans.* $17.

19. Sold flour at 2½% commission, invested ¾ of the amount of sales in sugar at 1½% commission; the balance, after deducting commissions, was $430. What was the flour sold for, and how much was invested in sugar?

The commission on sugar $= \frac{3}{4} \times \frac{4}{3} = 1\%$.

The whole commission on sugar and flour $= 3\frac{1}{2}\%$.

$\therefore \quad \frac{86}{100} = 430.$

$\frac{1}{100} = \frac{430}{86}.$

$\frac{100}{100} = 1 = 430 \times \frac{100}{86} = 500.$

3½% on the whole is 1¼% on ¼; the balance is $500, which is ¼ the sales of flour; therefore, flour sold for $2000 and $1500 invested in sugar.

20. Sold goods at 4% commission, and invested the proceeds, except my commissions, in other goods at 4% commission. My commissions were $200; what was the amount of sales and also of the investment?

$\frac{96}{100} \times \frac{104}{100} = \frac{96}{104}$ invested; $\frac{8}{104}$ = commission.

$\frac{8}{104} = \frac{1}{13} = 200.$

$\frac{13}{13} = 2600$, amount of sales.

158 PERCENTAGE.

21. Sold goods and invested the proceeds, retaining the commissions, which were 4% on the sales and 2% on the investment. The amount invested was $8000; what was the amount of sales and of the commissions?

$$\tfrac{94}{100} \times \tfrac{100}{102} = \tfrac{94}{102} = \tfrac{48}{51} = \tfrac{47}{50}, \text{ investment.}$$

$$\overset{500}{\cancel{8000}} \times \tfrac{17}{16} = 8500, \text{ sales.}$$
$$500, \text{ commissions.}$$

Or, $8000 \times \tfrac{102}{98} \times \tfrac{100}{102} = 8000 \times \tfrac{17}{16} = 8500.$

STOCKS, BONDS, Etc.

EXAMPLES.

1. What is the cost of 8 shares stock, $100 per share, at 5 per cent premium?

$$\$8\cancel{00} \times \tfrac{105}{100} = \$840.$$

2. What will 12 shares stock ($50) cost, at 15% discount? and what is the discount?
   Ans. $510 cost, $90 discount.

3. Bought $200 gold at ½% premium. What is the cost and premium?   *Ans.* $201 cost, $1 premium.

4. Bought a draft on New York for $500, premium ¼%. What is the cost and premium?
   Ans. $1¼ premium, $501¼ cost.

5. Sold $750 uncurrent funds, at 5% discount. What was the amount of sale, and the discount?
   Ans. $712½ sale, 37½ discount.

6. Bought 50 shares R.R. stock ($50) at 64%, and sold them at 69%. What did I gain?

$$\$50 \times 50 = \$25\cancel{00} \times \tfrac{5}{100} = \$125 \text{ gain.}$$

PERCENTAGE.

7. Bought stock at 105%, and sold it at 96%. What was lost on 60 shares ($50)? *Ans.* $270.

8. What is the difference in the cost of a draft on New York for $5000 at ¼% premium, and one on New Orleans for the same sum at ½% discount? *Ans.* 37½.

9. Purchased per broker, 20 shares stock ($50) at 15% discount, brokerage ½%. What was the cost?

$$\$1\cancel{000} \times \frac{\overset{17}{\cancel{34}\cancel{1}}}{\cancel{100}} \times \frac{\cancel{201}}{\cancel{200}} = \frac{3417}{4} = \$854\tfrac{1}{4}.$$

10. Bought 15 bonds (face $250) at 15% discount, brokerage ½%; sold them at 10% premium, brokerage ½%. What was the gain?

Bought 15 × 250 × $\frac{85}{100}$ × $\frac{201}{200}$.
Sold   15 × 250 × $\frac{110}{100}$ × $\frac{199}{200}$.

REM.—Brokerage has to be paid both on the purchase and sale; it will increase the cost, but diminish the sales.

PROBLEMS.

To find the rate per cent of gain or loss:

The ratio of the gain or loss reduced to hundredths, either as a common fraction or a decimal fraction, the number of hundredths will be the rate per cent.

EXAMPLES.

1. Paid $2525 for a draft on New York for $2500. What percentage was the premium?

$$\begin{array}{r}2525\\2500\\\hline 25\end{array}$$

$\dfrac{25}{2500}$ = rate of premium = $\tfrac{1}{100}$, or 1%.

2. Bonds (face $2500) cost $2550, brokerage $37½. What was the percentage of premium?

$$\begin{array}{r} 2550 \\ 37\tfrac{1}{2} \\ \hline 2512\tfrac{1}{2} \\ 2500 \\ \hline 12\tfrac{1}{2} \end{array}$$

$\frac{12\tfrac{1}{2}}{2500}$, rate of premium $= \frac{1}{100}$, or $\tfrac{1}{2}\%$, *Ans.*

3. Bonds (face $5000) cost $4275, brokerage $25. What per cent was the discount?

$$\begin{array}{rr} 4275 & 5000 \\ 25 & 4250 \\ \hline 4250 & 750 \end{array}$$

$\frac{750}{5000}$, rate of discount $= \frac{15}{100}$, or 15%, *Ans.*

4. Received $5.40 for a note of $6; what per cent was the discount?

$$\begin{array}{r} 6.00 \\ 5.40 \\ \hline .60 \end{array}$$

$\frac{.60}{6.00} = \frac{10}{100} = 10\%$.

5. Bought gold at 5% premium, the premium alone was $15; how much gold was bought?

$\frac{5}{100} = 15,$ $\frac{1}{100} = 3,$ $\frac{100}{100} = \$300.$

6. Sold bonds at 15% discount, and received $210 less than the face? What was the face value?

$\frac{15}{100} = 210;$ $\frac{100}{100} = 210 \times \frac{100}{15} = \$1400.$

PERCENTAGE.

7. Bought bonds at 60%, sold them at 80%; brokerage in the purchase and sale 1½%; the gain was $179. What was the face of the bonds?

$$1\tfrac{1}{2}\% \text{ of } 60\% = \tfrac{3}{2} \times \tfrac{3}{5} = \tfrac{9}{10}\%.$$
$$1\tfrac{1}{2}\% \text{ of } 80\% = \tfrac{3}{2} \times \tfrac{4}{5} = \tfrac{12}{10}\%.$$
$$\tfrac{21}{10}\% = 2\tfrac{1}{10}.$$

$2\tfrac{1}{10}\%$ is paid for brokerage, the gain is

$$20\% - 2\tfrac{1}{10}\% = 17\tfrac{9}{10}\%.$$
$$17\tfrac{9}{10} = \tfrac{179}{1000} = \$179.$$
$$\tfrac{1000}{1000} = 1 = 179 \times \tfrac{1000}{179} = \$1000, \textit{Ans.}$$

8. What is the face of a draft costing $3193.19, bought at a premium of 1½%?

$$\$3193.19 \times \tfrac{200}{203} = \$3146.$$

9. Bought a draft on New Orleans at 1 per cent discount for $990; what was the face of the draft?

Ans. $1000.

10. Invested $6750 in stocks at 25% discount. What was the par value of the stocks? *Ans.* $9000.

11. Invested $6250 in stocks at 60%, and brokerage 2½% on the par value. What is the par value of the stocks? *Ans.* $10000.

12. How much gold at 5½% premium, can be bought for $2268.25? *Ans.* $2150.

13. Bought goods at 80 cts. per yard, and sold at 90 cts. per yard. What per cent did I gain?

$$\tfrac{10}{80} = \tfrac{50}{400} = \tfrac{12\tfrac{1}{2}}{100} = 12\tfrac{1}{2}\%.$$

14. Bought goods at 80 cts., and sold at 70. What per cent did I lose?

$$\frac{10}{80} = \frac{12\frac{1}{2}}{100} = 12\frac{1}{2}\%.$$

15. Bought goods for $1 and sold for $4. What per cent did I gain?

$$\frac{3}{1} = \frac{300}{100} = 300\%.$$

16. Bought goods for $4 and sold for $3. What per cent did I lose?

$$\frac{1}{4} = \frac{25}{100} = 25\%.$$

17. Bought goods for $5 and sold for $2. What per cent did I lose?

$$\frac{3}{5} = \frac{60}{100} = 60\%.$$

18. Bought for $1 and sold for 75 cts. What per cent did I lose?

$$\frac{25}{100} = 25\%.$$

19. Bought goods for $160 and sold for $180. What per cent did I gain?

$$\frac{20}{160} = \frac{1}{8} = \frac{25}{200} = \frac{12\frac{1}{2}}{100} = 12\frac{1}{2}\%.$$

INSURANCE.

Insurance of property is a guarantee of a certain sum of money in case the property is lost by fire or any casualty.

The contract for insurance is termed a ***Policy***, and the sum paid annually is the ***Premium***.

PERCENTAGE. 163

The *Premium* is a certain percentage on the amount of risk.

There are different modes of insurance, but all are dependent on the principle of percentage.

EXAMPLES.

1. What is the premium on property on which a risk of $6000 is taken, at $\frac{1}{2}$% for one year?

$$\$6000 \times \frac{\frac{1}{2}}{100} = \frac{1}{200} = \$30.00.$$

2. What is the premium on a risk of $5000, on which a premium note is given for 5% of the risk, and the interest on the note is 4%?

Note $= 5000 \times \frac{5}{100} = \$250.$
Interest on note $= 250 \times \frac{4}{100} = \$10 =$ Premium.

If the policy cost $1.00, the expense for the first year is $11.00.

REM.—Sometimes the premium is made to cover both property and premium, in which case the risk is equal to the sum of the value on property and the premium.

3. What amt. must I have insured at $\frac{1}{2}$% prem., to cover property and prem., when the risk on the property is $9950?

As the premium is $\frac{1}{2}$%
The property must be $99\frac{1}{2}$%
As the whole is   100%.

∴ $\frac{99\frac{1}{2}}{100} = \frac{199}{200} =$ Property.

$\frac{199}{200} = 9950.$
$\frac{1}{200} = 50.$
$\frac{200}{200} =$ the whole risk $= 10000.$

PERCENTAGE.

INTEREST.

Interest is an allowance for the use of money. It is reckoned by percentage; thus, 5%, 6%, etc., meaning for a year, when not otherwise expressed; for any other time it is as the ratio of the time; thus, the interest of $100 at 6% is $6 for a year, for two years $12, and for six months $3.

PROBLEMS.

1. Find the interest of $150, at 6%, for 1 year.

$$\$150 \times \tfrac{6}{100} = \$9.$$

For 8 months.

$$\$1.50 \times \tfrac{6}{100} \times \tfrac{\overset{4}{8}}{\underset{2}{12}} = 1.50 \times \tfrac{4}{100} = \$6.00.$$

For 6 months.

$$\$1.50 \times \tfrac{6}{100} \times \tfrac{\overset{3}{6}}{\underset{2}{12}} = \$4.50.$$

For 14 months.

$$\$1.50 \times \tfrac{6}{100} \times \tfrac{\overset{7}{14}}{\underset{2}{12}} = 1.50 \times \tfrac{7}{100} = \$10.50.$$

Cor.—At 6%, the rate per cent. for any number of months is ½ the number of months; thus, for 8 months it is 4%, for 6 months it is 3%, and for 14 months 7%.

2. Find the interest of $150, at 6%, for 129 days?

$$\$150 \times \tfrac{6}{100} \times \tfrac{129}{360} = \$150 \times \tfrac{129}{6000} = \$3.225.$$

Cor.—The interest of a sum of money for any number of days is equal to the product of the sum of money and the number of days divided by 6000; or, if the number of dollars be multiplied by the number of days and this product divided by 6, the quotient is the interest in mills; point off three decimals and it is reduced to dollars, cents, and mills.

If the rate of interest is 7%, add $\frac{1}{6}$; if 8%, add $\frac{1}{3}$; if 9%, add $\frac{1}{2}$; if 5%, deduct $\frac{1}{6}$; if 4%, deduct $\frac{1}{3}$; if 3%, take $\frac{1}{2}$.

The rate for 200 months is 100%; that is, the interest is equal to the principal.

200 months of $100 is	$100.
20 months of $100 is	$10.
2 months of $100 is	$1.
30 days, or 1 month	$0.50.
3 days of $100 is	$.05.
1 day of $100 is	$.01⅔.
2 days of $100 is	$.03⅓.

EXAMPLES.

1. Find the interest of $100 for 1 year, at 6%, 7%, 8%, 9%, 10%, 5%, 4%, 3%, and 2%.

For 1 year at 6% it is	$6.00
	1.00
" 7% add $\frac{1}{6}$ =	$7.00
" 8% add $\frac{1}{3}$, $6 + $2 =	8.00
" 9% add $\frac{1}{2}$, $6 + $3 =	9.00
" 10% = $\frac{1}{10}$; ∴ $\frac{100}{10}$ =	10.00
" 5% deduct $\frac{1}{6}$, $6 − $1 =	5.00
" 4% deduct $\frac{1}{3}$, $6 − $2 =	4.00
" 3% take $\frac{1}{2}$; $\frac{6}{2}$ =	3.00
" 2% take $\frac{1}{3}$; $\frac{6}{3}$ =	2.00

2. Find the interest of $625 at 6% for 8 months. Rate for 8 mo. is $\frac{4}{100}$ or 4%.

$$625 \times \tfrac{4}{100} = \$25.00.$$

For 8 mo. and 20 days $= 8\tfrac{2}{3}$, $\tfrac{1}{3}$ of $8\tfrac{2}{3} = 4\tfrac{1}{3}$.

$$\therefore 625 \times \frac{4\tfrac{1}{3}}{100} = 625 \times \tfrac{13}{300} = \$27.08\tfrac{1}{3}.$$

If months and days are computed separately,

$$625 \times \tfrac{4}{100} = \$25.00.$$
$$625 \times \tfrac{2}{100} \times \tfrac{1}{300} = \underline{\;\;2.08\tfrac{1}{3}\;\;}$$
$$\$27.08\tfrac{1}{3}$$

3. What is the interest of $845 at 6% for 1 year 6 months and 24 days?

$18\tfrac{1}{4}$ mo. Per cent. $= \dfrac{9\tfrac{2}{3}}{100}$ $\dfrac{9\tfrac{2}{3}}{100} = \tfrac{47}{500}.$

$$\therefore \$845 \times \tfrac{47}{500} = \$79.43.$$

Or reduce to days 1 year $= 360$
Reduce to days $\tfrac{1}{2}$ year $= 180$
$\phantom{\text{Reduce to days }\tfrac{1}{2}\text{ year} = }\underline{24}$
$\phantom{\text{Reduce to days }\tfrac{1}{2}\text{ year} = }564$

$$\therefore 845 \times \tfrac{47}{1000} = 845 \times \tfrac{47}{500} = \text{same as above.}$$

4. What is the interest of $845 at 7% for 1 year 6 months and 24 days? at 8%? 5%? 4%?

At 6% = 79.43   At 6% = $79.43
Add $\tfrac{1}{3}$ = $\underline{\;\;26.47\tfrac{2}{3}\;\;}$ Add $\tfrac{1}{6}$ = $\underline{\;\;13.23\tfrac{5}{6}\;\;}$
At 8% = 105.90$\tfrac{2}{3}$ At 7% = 92.66\tfrac{5}{6}$

PERCENTAGE. 167

At 6% = 6) 79.43   At 6% = $79.43
Deduct ⅙ = 13.23½   Deduct ⅓ = 26.47⅔
At 5% = 66.19¼   At 4% = $52.95⅓

5. What is the interest of $648 at 6% for 3 years 5 months and 18 days?

3 years = 1080 days.
5 months = 150 days.
18 days.
1248 days.

$$\$648 \times \frac{\cancel{1248}^{208}}{\cancel{6000}} = \$134.784.$$
$$1000$$

REM.—When the denominator is a divisor and is reduced to 10, 100, 1000, etc., it is best not to cancel further, as it is so convenient in use.

6. What is the interest of $540 from Oct. 10th, 1872, to Aug. 9th, 1875, at 6%?

2 years = 720 days.
9 months = 270 days.
29 days.
1019

	yrs.	mos.	days.
	1875	8	9
	1872	10	10
	2	9	29

$$\cancel{\$540} \times \frac{\cancel{1019}^{9}}{\cancel{600}} = \$91.71$$
$$\phantom{\cancel{\$540} \times }100$$

7. What is the interest of $741 at 7% for 1 year 8 months and 20 days? 20 months and 20 days at 6%?

1 year = 360 days.
8 months = 240 days.
20 days.
620 days.

$$\overset{247}{\cancel{\$741}} \times \frac{\overset{31}{\cancel{620}}}{\underset{\underset{100}{\cancel{300}}}{\cancel{6000}}} = \$76.57.$$

\$741 for 20 months	=	74.10.	Divide by 30 for da.
\$741 for 20 days	=	2.47	
6%	=	76.57	
Add $\frac{1}{6}$,		12.76$\frac{1}{6}$	
7%	=	\$89.33$\frac{1}{6}$	

8. What is the interest of \$541.64 for 2 years 6 months and 15 days at 7%.

2 years	=	720 days.
6 months	=	180 days.
		15 days.
		915 days.

$$\overset{270.82}{\cancel{\$541.64}} \times \frac{\overset{305}{\cancel{915}}}{\underset{\underset{1000}{\cancel{2000}}}{\cancel{6000}}} = \frac{270.82 \times 305}{1000} = \text{Int. at } 6\%.$$

Add $\frac{1}{6}$.

9. What is the interest of \$640.74 from April 12th, 1864, to Feb. 22d, 1871, at 9%?

yr.	mo.	da.
1871	2	22
1864	4	12
6	10	10

6 years	=	2160 days.
10 months	=	300 days.
		10 days.
		2470 days.

$$\overset{106.79}{\$640.74} \times \frac{\cancel{2470}}{\underset{100}{\cancel{6000}}} = \$263.7713 = \text{Int. at } 6\%.$$

To which add $\frac{1}{2}$	=	131.8856
Interest at 9%	=	\$395.6569

PERCENTAGE.

10. A note was given Apr. 1st, 1872, for $6000, bearing interest from date at 6%. On the back of the note were the following credits: May 1st, 1873, $1000; June 1st, 1874, $1000; July 1st, 1875, $2000. What sum was still due August 1st, 1877?

REM.—When there are partial payments on a note, and whenever the payment is greater than the interest, it is computed until the time of the payment and added to the principal, from which the payment is deducted, and the balance is then regarded as the principal of the note.

When the payment is less than the interest, the interest must be computed until all the payments are greater than the interest, and then add interest and principal, and from their sum deduct the sum of all the payments, and the balance is then regarded as the principal of the note.

$$
\begin{array}{r}
6000 \\
.06\tfrac{1}{4} \\ \hline
30.00 \\
360.00 \\
6390 \\
1000 \\ \hline
5390, \text{ Prin.} \\
.06\tfrac{1}{4} \\ \hline
26.95 \\
323.40 \\
350.35 \\
5390 \\ \hline
5740.35 \\
1000 \\ \hline
4740.35, \text{ Prin.} \\
.06\tfrac{1}{4} \\ \hline
2370.175 \\
28442.10 \\
308.12275 \\
4740.35 \\ \hline
5048.47275 \\
2000 \\ \hline
3048.47275, \text{ Prin.} \\
.12\tfrac{1}{2} \\ \hline
15.24236375 \\
365.8167300 \\ \hline
381.06 \\
3048.47 \\ \hline
\end{array}
$$

Due August 1st, 1877 . . . . $3429.53

$5640.50.   April 1st, 1870.

11. Six months after date I promise to pay to John Johnson or order, five thousand six hundred and forty $\frac{50}{100}$ dollars, value received.   P. PAYSON.

This note is credited as follows:
    1870 Oct. 1st, forty $\frac{50}{100}$ dollars.
    1871 Oct. 1st, one hundred dollars.
    1872 Oct. 1st, two hundred dollars.
    1873 Oct. 1st, one thousand dollars.

What sum is due Oct. 1st, 1875?

The first payment is made when there is no interest. As the other payments do not equal the interest until Oct. 1st, 1873, the interest is computed until that time.

The note is due Oct. 1st, 1870, when there is a payment which must be deducted.

    Interest to Oct. 1st, 1873, 3 years, $5600 \times \frac{18}{100}$.
        Oct. 1st, 1873, sum of payments.
    Interest to Oct. 1st, 1875, $5308 \times \frac{12}{100}$.

$5640.50
  40.50
―――――――
5600 Principal.
1008 Int. to 1st Oct., 1873.
―――
6608 Amt.   "   "
1300 Sum paid "   "
―――
5308 Prin. 1st Oct., 1873.
 639.96 Int. to 1st Oct., 1875.

Amt. due Oct. 1st, 1875 = $5947.96

REM.—When interest is not inserted in the note it takes the legal rate, which is generally 6%. In all places the years and months are computed alike, but in New York and several other States, the days are in their ratio to 365 instead of 360.

REVIEW.

PERCENTAGE.

ORAL QUESTIONS.

1. One is what part of 100, or 1 is what per cent of 100? *Ans.* $\frac{1}{100}$, or 1%.
2. 2 is what part of 100, or 2 is what per cent of 100? *Ans.* $\frac{1}{50}$, or 2%.
3. 3 is what part of 100, or 3 is what per cent of 100? *Ans.* $\frac{3}{100}$, or 3%.
4. 20 is what part of 100, or 20 is what per cent of 100? *Ans.* $\frac{1}{5}$, or 20%.
5. 25 is what part of 100, or 25 is what per cent of 100? *Ans.* $\frac{1}{4}$, or 25%.
6. 50 is what part of 100, or 50 is what per cent of 100? *Ans.* $\frac{1}{2}$, or 50%.
7. $\frac{1}{3}$ is what % of 1? *Ans.* $33\frac{1}{3}$%.
8. $\frac{2}{3}$ is what % of 1? *Ans.* $66\frac{2}{3}$%.
9. $\frac{1}{7}$ is what % of 1? *Ans.* $14\frac{2}{7}$%.
10. $\frac{2}{7}$ is what % of 1? *Ans.* $28\frac{4}{7}$%.
11. 5 is what per cent of 10? *Ans.* 50%.
12. 10 is what per cent of 10? *Ans.* 100%.
13. 3 is what per cent of 10? *Ans.* 30%.
14. $2\frac{1}{2}$ is what per cent of 10? *Ans.* 25%.
15. 2 is what per cent of 10? *Ans.* 20%.
16. 1 is what per cent of 10? *Ans.* 10%.
17. $\frac{1}{2}$ is what per cent of 10? *Ans.* 5%.

PERCENTAGE.

18. ¼ is what per cent of 10? $Ans.\ 2\frac{1}{2}\%$.
19. 1 is what % of 6? $Ans.\ \frac{1}{6} = 16\frac{2}{3}\%$.
20. 2 is what % of 6? $Ans.\ \frac{1}{3} = 33\frac{1}{3}\%$.
21. 3 is what % of 6? $Ans.\ \frac{1}{2} = 50\%$.
22. 4 is what % of 6? $Ans.\ \frac{2}{3} = 66\frac{2}{3}\%$.
23. 5 is what % of 6? $Ans.\ \frac{5}{6} = 83\frac{1}{3}\%$.
24. 6 is what % of 6? $Ans.\ \frac{6}{6} = 100\%$.
25. ½ is what % of 6? $Ans.\ \frac{1}{12} = 8\frac{1}{3}\%$.
26. ¼ is what % of 6? $Ans.\ \frac{1}{24} = 4\frac{1}{6}\%$.

INTEREST.

1. The interest of $100 at 6% for 12 months is
$$\$100 \times \tfrac{6}{100} \times \tfrac{12}{12} = \$6.00.$$

2. The interest of $100 at 6% for 6 months is
$$\$100 \times \tfrac{6}{100} \times \tfrac{6}{12} = \$3.00.$$

1st PRINCIPLE.—The interest is equal to the product of the principal, rate, and time; that is, the interest is the product of three factors.

2D PRINCIPLE.—Any one of these factors may be found by dividing the product of the three factors by the product of the other two.

COR. 1.—If the interest is divided by the product of the principal and rate, the quotient is the time; thus,
$$\frac{6}{100 \times \tfrac{6}{100}} = \tfrac{6}{6} = 1 \text{ year} = \text{time}.$$

COR. 2.—If the interest is divided by the product

PERCENTAGE.

of the principal and time, the quotient is the rate; thus,

$$\frac{6}{100 \times 1\frac{2}{8}} = \frac{8}{100} = \text{rate}.$$

COR. 3.—If the interest is divided by the product f the rate and time, the quotient is the principal; thus,

$$\frac{6}{\frac{8}{100} \times 1\frac{2}{8}} = 6 \times \frac{100}{8} = \$100 = \text{principal}.$$

PROBLEMS.

1. What is the interest of $540 at 6% for 3 yr. 4 mo. 20 da.

   Ans. $540 \times \frac{8}{100} \times \frac{1220}{360} = \$109.80$, Interest for 3 yr. 4 mo. 20 da.

2. The principal is $540, the rate 6%, and the interest $109.80; what is the time?

   Ans. $\dfrac{\$109.80}{540 \times \frac{8}{100}} = \dfrac{109.80}{32.40} = 3\frac{7}{18} = 3$ yr. 4 mo. 20 da.

3. The principal is $540, the time $3\frac{7}{18}$ yr., and the interest $109.80; what is the rate?

   Ans. $\dfrac{109.80}{540 \times 3\frac{7}{18}} = \dfrac{109.80}{540 \times \frac{61}{18}} = \dfrac{109.8}{30 \times 61} = \dfrac{109.8}{1830.0} = \frac{183}{3100} = \frac{8}{100}$, rate.

4. The time is $3\frac{7}{18}$ yr., the rate 6%, and the interest $109.80; what is the principal?

   Ans. $\dfrac{\$109.80}{\frac{61}{18} \times \frac{8}{100}} = \dfrac{109.80}{\frac{61}{300}} = \overset{1.80}{\cancel{109.80}} \times \frac{300}{61} = \$540.00$.

PERCENTAGE.

EXAMPLES.

1. The principal is $1250, the rate 6%, and the interest $173.75; what is the time?
 Ans. 2 yr. 3 mo. 24 da.
2. The principal is $625, the time $2\frac{11}{16}$ years, and the interest $86.87½; what is the rate? *Ans.* 6%.
3. The time is $2\frac{11}{16}$ years, the rate 6%, and the interest $260.62½; what is the principal? *Ans.* $1875.00.
4. The principal is $600, the rate 7%, and the interest $57.75; what is the time? *Ans.* 1 yr. 4 mo. 15 da.
5. The principal is $600, the rate 8%, and the interest $105.60; what is the time? *Ans.* 2 yr. 2 mo. 12 da.
6. The principal is $600, the time 1¾ years, and the interest $57.75; what is the rate? *Ans.* 7%.
7. The principal is $600, the time 2¼ years, and the interest $105.60; what is the rate? *Ans.* 8%.
8. The time is 1¾ years, the rate 7%, and the interest $57.75; what is the principal? *Ans.* $600.
9. The time is 2¼ years, the rate 8%, and the interest $105.60; what is the principal? *Ans.* $600.
10. What is the interest of $6000 at 9% for 5 yr. 6 mo. 18 da.? *Ans.* $2997.
11. What is the interest of $6000 at 6% for 5 yr. 6 mo. 18 da.? *Ans.* $1998.
12. What is the interest of $6000 at 3% for 5 yr. 6 mo. 18 da.? *Ans.* $999.
13. What is the interest of $6000 at 4% for 5 yr. 6 mo. 18 da.? *Ans.* $1332.
14. What is the interest of $6000 at 5% for 5 yr. 6 mo. 18 da.? *Ans.* $1665.

PERCENTAGE.

REM.—In the computation of interest we find the interest equal to the product of the principal, rate per cent, and time. By using the initial letters for these, we get convenient formulas; thus,

Let P = principal, r = rate per cent, t = time, and I = interest; we have the equation

(1.) $\qquad P \times r \times t = I.$

Divide both members by $r \times t$,

(2.) $\qquad P = \dfrac{I}{r \times t}.$

Principal equal to interest divided by the product of rate and time.

(3.) $\qquad r = \dfrac{I}{P \times t}.$

Rate equal to interest divided by the product of principal and time.

(4.) $\qquad t = \dfrac{I}{P \times r}.$

Time equal to interest divided by the product of principal and rate.

ENUNCIATION OF THE FORMULAS.

1. The interest of a sum of money is equal to the sum, called principal, multiplied by the rate and the time.

2. The principal is equal to the interest divided by the product of the rate and the time.

3. The rate is equal to the interest divided by the product of the principal and the time, and then reduced to hundredths.

PERCENTAGE.

4. The time is equal to the interest divided by the product of the principal and the rate.

The four formulas enable us to determine any one of the four quantities when the other three are known.

PROBLEMS.

1st FORMULA.—The principal is $750, the rate 6%, time 8 months. What is the interest?

$$I = \$750 \times \tfrac{6}{100} \times \tfrac{8}{12} = \$30.00 = \text{Interest.}$$

2D FORMULA.—The interest is $30, the rate 6%, and the time 8 months. What is the principal?

$$P = \frac{\$30}{\tfrac{6}{100} \times \tfrac{8}{12}} = \frac{30}{\tfrac{4}{100}} = 30 \times \tfrac{100}{4} = \$750, \text{Principal.}$$

3D FORMULA.—The interest is $30, the principal $750, and the time 8 months. What is the rate?

$$r = \frac{\$30}{750 \times \tfrac{8}{12}} = \tfrac{30}{500} = \tfrac{6}{100} = 6\%, \text{Rate.}$$

4TH FORMULA.

$$t = \frac{30}{750 \times \tfrac{6}{100}} = \tfrac{30}{45} = \tfrac{2}{3} = \tfrac{8}{12}, \text{Time.}$$

EXAMPLES.

1. Compute the interest on $5464.50, at 5%, for 1 year 9 months and 18 days.   *Ans.* $491.805.

2. Compute the interest on $5464.50, at 8%, for 1 year 9 months and 18 days.   *Ans.* $786.888.

PERCENTAGE.

3. Compute the interest on $5464.50, at 10%, for 1 year 9 months and 18 days?

4. The interest is $630, the rate 6%, and the time 18 months. What is the principal? *Ans.* $7000.

5. The interest is $540, the rate 9%, and the time 24 months. What is the principal? *Ans.* $3000.

6. The interest is $600, the rate 5%, and the time 30 months. What is the principal?

7. The interest is $300, the principal $2400, and the time 30 months. What is the rate? *Ans.* 5%.

8. The interest is $560, the principal $3000, and the time 32 months. What is the rate? *Ans.* 7%.

9. The interest is $140, the principal $1500, and the time 16 months. What is the rate?

10. The interest is $270, the principal $3000, and the rate 6%. What is the time? *Ans.* $1\frac{1}{2}$ years.

11. The interest is $360, the principal $2400, and the rate 8%. What is the time? *Ans.* $22\frac{1}{2}$ months.

12. The interest is $66, the principal $500, and the rate 6%. What is the time?

13. The principal is $440, the interest $88, and the time 4 years. What is the rate? *Ans.* 5%.

14. The principal is $650, the interest $78, and the rate 6%. What is the time? *Ans.* 2 years.

15. The interest is $48, the rate 4%, and the time 3 years. What is the principal? *Ans.* $400.

16. In what time will any sum of money double itself at 6% simple interest?

$$t = \frac{I}{P \times r} = \frac{P}{P \times r} = \frac{1}{r} = \frac{1}{\frac{6}{100}} = 1 \times \frac{100}{6} = 16\frac{2}{3} \text{ years.}$$

REM.—When the principal is doubled, the interest is equal to the principal; that is, $I = P$.

BANKING.

Bank Discount is reckoned on the face of the note the same as interest.

It is called discount, as the interest for the time the note is given and three days grace is deducted from the face of the note, and the borrower receives the difference. A note in bank is not considered due until three days after the time specified.

The bank discount on a rate for

$100 at 60 days = $\cancel{100} \times \frac{1}{\cancel{100}} \times \frac{\cancel{63}}{\cancel{360}} = \frac{21}{20} = \$1.05$.

Proceeds $100 − $1.05 = $98.95.

$100 at 90 days = $\cancel{100} \times \frac{1}{\cancel{100}} \times \frac{\cancel{93}}{\cancel{360}} = \frac{31}{20} = \$1.55$.

Proceeds $100 − $1.55 = $98.45.

$500 at 90 days = $\cancel{500} \times \frac{1}{\cancel{100}} \times \frac{\cancel{93}}{\cancel{360}} = \frac{31}{4} = \$7.75$.

Proceeds $500 − $7.75 = $492.25.

$324 at 90 days = $\cancel{324} \times \frac{\cancel{93}}{\cancel{1000}} = \frac{5021}{1000} = \$5.022$.

Produce $324 − $5.02 = $318.98.

A bank account is closed at the end of the year, and the next year is begun by bringing forward the balance which belongs to the credit side of account, as overdrawing is not permitted.

To this balance each deposit is added at the date on which it is made, and from the sum on the credit side is subtracted each sum drawn by check at its date.

Each sum or balance is multiplied by the number of days from its date until the next transaction; lastly, the difference at the last transaction by the time until the end of the year.

The sum of all these products will be the number of days that one dollar is at interest.

The last sum or difference on the credit side will be the principal.

DR.   JAMES KENNEDY in acct. with MERCHANTS' BANK.   CR.

1877.				1876.			DA.	PRO.
Jan. 6.	To check.	$200		Dec. 31.	By bal. old acct.	$600	6	3600
" 15.	" "	100		Jan. 10.	By cash $300.	400	4	1600
" 18.	" "	200		Jan. 25.	By cash $750.	700	5	3500
" 29.	" "	350				600	3	1800
						400	7	2800
						1150	4	4600
					Balance items.	800	2	1600
					Bal. interest.		3.25	6)19,500
					Total bal.	$803.25		Int. 3.25

This account is rendered for one month. The debit side shows the sums drawn and their dates, the column of products gives the number of days that one dollar is on interest. The balances show the sum of money in bank after each transaction. Each month's account is kept in the same manner, and at the end of the year the sum of the last balance and the interest is the total balance.

REM.—If a creditor were permitted to overdraw, interest on the balance would be computed on the debit side, and the difference between the two columns of products would belong to the side having the greater sum.

Balance the following account:

DR.						E. F. BENTON in account with L. R. SANDS.				CR.
1878.					1877.					
Jan. 10.	To check.	$105			Dec. 31.	By bal. old acct.	$185	10	1850
Jan. 15.	" "	200					80	5	400
Jan. 31.		120	10	1200	Jan. 25.	By cash $890.	190	6	1050
							.35		
						Total balance	$180.85		1200
							6) 2.130		
							Int. .35		

TRUE DISCOUNT

is the abatement on notes not yet due, or the difference between the sum of money at a certain rate, whose amount will be equal to the face of the note when due. Thus, a note for $100 due in one year without interest when money is worth 6%, is such a sum as will amount to $100 in one year.

As $100 now is worth $106 at the end of the year, the present worth of money due in one year is $\frac{100}{106}$ of the money due in one year.

$112 at the end of two years is worth $100 now.
$118 at the end of three years is worth $100 now, etc.

COR.—The ratio for discounting is a diminishing ratio having 100 for the numerator, and the denominator is 100 increased by the interest for the time and rate.

EXAMPLES.

1. What is the present worth of a note due in 1 year for $324, money worth 6%?

$$324 \times \frac{100}{106} = \$305.66 +$$

PERCENTAGE.

2. What is the present worth of a note due in 2½ years for $675 at 5%? *Ans.* $600.

3. What is the present worth of a note due in 4 years for $960 at 5%? *Ans.* $800.

4. What is the present worth of a note due in 15 months for $445 at 9%? *Ans.* $400.

5. What is the present worth of a note due in 9 months for $721 at 4%? *Ans.* $700.

6. What is the present worth of a note due in 13 months for $678 at 12%? *Ans.* $600.

EXCHANGE.

Exchange is the system by which payments are made at a distant place by means of Bills of Exchange or Drafts.

In a Bill of Exchange or draft, the person giving it or signing it is called the **Drawer** or **Maker;** the one to whom it is addressed is the **Drawee**, and the person to whom it is ordered to be paid is the **Payee.**

The person in possession of the draft is the **Holder,** and if he sells it he must endorse it, which makes him responsible for the payment, unless otherwise specified.

Exchange between the different cities of our own country is **Domestic**, and that with a foreign country is called **Foreign Exchange.**

Exchange may be direct or circuitous; the latter is sometimes found to be advantageous.

DOMESTIC EXCHANGE.

PROBLEM.

To find the cost of a draft on Philadelphia or New York, or any distant city, when at a premium and when at a discount; thus, if the premium is 1%, the draft will cost $\frac{101}{100}$ of its face, which is the sum to be paid; when at a discount of 1%, it will cost $\frac{99}{100}$ of its face.

EXAMPLES.

1. What is the cost of a draft on Philadelphia for $3024, at ½% premium?

$$3024 \times \frac{100\frac{1}{2}}{100} = \frac{6079\frac{1}{4}}{2} = \$3039.12.$$

2. What is the cost of a draft on New Orleans for $5000, at ½% discount?

$$\frac{1}{2}\% \text{ of } \$5000 = \$25.$$
$$\$5000 - 25 = \$4975.$$

3. What will this draft cost if drawn at thirty days, interest at 6%? 3 days grace must be allowed.

$$\frac{33}{60} = \frac{30}{60} + \frac{3}{60} = \frac{33}{60} = \text{rate per cent;}$$
$$5000 \times \frac{33}{60} = \$52.50.$$

The interest and discount must be reckoned on the face.

$$\$5000 - \$52.50 = \$4947.50.$$

4. My agent had in his possession $3288.60, which I directed him to remit; the draft received was $3240; what was the rate of exchange?

PERCENTAGE.

$$3288.60$$
$$\underline{3240.00}$$
$$\$48.60 = \text{paid for a draft of } \$3240.$$

The rate is $\dfrac{48.60}{3240.00}$, reduce by $32.4 = \dfrac{1\frac{1}{2}}{100} = 1\frac{1}{2}\%$; the result will be the same if reduced to a decimal; thus,

$$\tfrac{1.5}{100} = 100\,)\,1.5000\,(\,.015$$

Rem.—The ratio of the face of the draft and the premium paid on it, reduced to a fraction whose denominator is 100, will be the rate per cent of the draft; the same result is obtained by reducing the fraction to a decimal. The premium is the numerator, and the amount on the face of the draft the denominator of the fraction.

5. What is the cost of a 60 day draft on New York for $5480, at ½% premium, interest off at 6%?

$$\text{Discount} = \tfrac{6\,6}{8}\% = \tfrac{5\,1}{2}\%.$$
$$\text{Premium} = \tfrac{1}{2}\% = \tfrac{4\,8}{8}\%.$$
$$\text{Discount above premium} = \tfrac{4\,4}{8}\%.$$

$$1\% \text{ of } \$5480 = 54.80.$$

½% of $5480	=	27.40	$5480.00
$\tfrac{1}{16}$% of $5480	=	2.74	30.14
Discount	=	$30.14	$5449.86, Cost.

6. What is the face of a draft which cost $1007.50, at ¾% premium?

7. What is the face of a draft that cost $992.50, at ¾% discount?

8. What is the face of a draft on which $5 premium is paid, and the rate 1%? $\tfrac{1}{100} = \$5$;
$$\tfrac{100}{100} = \$500, \textit{Ans}.$$

FOREIGN EXCHANGE.

EXAMPLES.

1. What is the cost in New York of a draft on London for £546 15s. 6d., at 8% premium?

$$12 \overline{)6d.} \qquad 15s.\ 6d.$$
$$20 \overline{)15.5} \qquad \underline{12}$$
$$.775 \qquad \tfrac{186}{240} = \tfrac{31}{40} = .775.$$

$$£546.775 \times \tfrac{8}{5} \times \tfrac{108}{100} = 109.355 \times 8 \times 3 = \$2624.52.$$

$$546\tfrac{31}{40} = \tfrac{21871}{40} \times \tfrac{8}{5} \times \tfrac{108}{100} = \$2624.52.$$

2. What amount of debt in London may be paid by depositing $5000 in New York, rate of exchange 8% in favor of London?

$$\$5000 \times \tfrac{1}{24} \times \tfrac{100}{108} = 125 \times \tfrac{25}{3} = £1041\ 13s.\ 4d.$$

3. What is the cost of a draft on London for £546.775 in New York, exchange $4.88 to the £?

$$546.775 \times 4.88 = \$2668.262.$$

4. What will a draft on London for £1041 13s. 4d. cost in New York, exchange $4.80 to the £?

PERCENTAGE. 185

$$240 \overline{\smash{)}\begin{array}{r}13s. \quad 4d. \\ 12 \\ \hline 16\cancel{0} \\ \hline 24\cancel{0}\end{array}} = \tfrac{2}{3}.$$

$$£1041\tfrac{2}{3} \times 4.80 = \$5000.$$

5. What is the value in Paris of $5000 in New York exchange $4.85 = £1, and £1 = 25.20 francs?

$$\frac{\overset{1000}{\cancel{\$5000}}}{\underset{\cdot 97}{\cancel{\$4.85}}} \times 25.20 = \underset{97}{\overset{1000}{252000}}.$$

6. What sum in New York will pay 25979$\tfrac{17}{97}$ francs in Paris, exchange as in 5th?

$$25979\tfrac{17}{97} \times \tfrac{485}{2520}.$$

7. What is the value in Hamburg of $5000 in New York, when $4.80 = £1, £1 = 25.20 francs, and 1.80 francs = 1 mark banco?

$$\overset{1250}{\cancel{\$5000}} \times \overset{7}{\underset{12}{\cancel{\tfrac{252}{48}}}} \times \underset{2}{\overset{5}{\cancel{\tfrac{18}{18}}}} = \tfrac{1250\times 7\times 5}{3} = 14583\tfrac{1}{3} \text{ mark banco.}$$

Machine

8. What sum in New York will pay 14583$\tfrac{1}{3}$ mark banco in Hamburg, exchange as above?

$$14583\tfrac{1}{3} \times \tfrac{18}{10} \times \tfrac{48}{252} = \$5000.$$

9. What debt in St. Petersburg will $5000 in New York pay, when $4.80 = £1 in London, £1 = 25.20f. in

186   *PERCENTAGE.*

Paris, 4f. = 1 thaler in Dantzic, and 1 thaler = 2.60 roubles?

$$\cancel{\$5000} \times \cancel{\tfrac{135}{2.56}}^{125} \times \cancel{\tfrac{1}{4}}^{21} \times \cancel{\tfrac{13}{1}}^{13} = \tfrac{125 \times 21 \times 13}{2} = 17062\tfrac{1}{2} \text{ roubles.}$$

These problems may be solved by compound equations and ratios.

10. What sum in New York will pay $17062\tfrac{1}{2}$ roubles in St. Petersburg, exchange as above?

$$17062.5 \times \frac{1}{\cancel{2.6}_{13}} \times \frac{4}{\cancel{25.2}_{21}^{12}} \times \frac{\cancel{4.0}^{\;8}_{\;11}}{1} =$$

REM.—This circuitous system of Foreign Exchange has led us into the same thing in our own country when it is found to be advantageous; thus,

11. A merchant in St. Louis has a bill of $5000 to pay in New York; in St. Louis a draft on New York costs $1\tfrac{1}{2}\%$ premium, whilst New Orleans funds can be bought at $1\tfrac{1}{4}\%$ premium, and at New Orleans Havana funds are $\tfrac{1}{2}\%$ discount; and at Havana, New York funds are at a discount of $1\tfrac{1}{4}\%$. What sum in St. Louis will pay the bill by the direct and by the circuitous method?

Direct, $\cancel{\$5000}^{25} \times \dfrac{\cancel{101\tfrac{1}{2}}}{\cancel{100}} \times \cancel{\tfrac{203}{200}} = \$5075.$

And $\cancel{\$5000}^{25} \times \cancel{\tfrac{79}{100}}_{16} \times \cancel{\tfrac{101}{100}} \times \cancel{\tfrac{101}{100}}_{\;16}^{\;81} = \$4974.22\tfrac{11}{12}.$

PERCENTAGE.

12. What is $4974.22\frac{17}{44}$ in St. Louis worth in New York by the circuitous method?

$$4974.22\tfrac{17}{44} \times \tfrac{100}{101} \times \tfrac{100}{102} \times \tfrac{101}{100} = \$5000.$$

I have not given a table of foreign values, as they are not permanent. The principle is the same in all cases.

REM. 1.—In the ratios, see that the decimals in the numerator and denominator are equal, and then they are like quantities and can be used as such.

REM. 2.—When any thing is purchased at a premium, the ratio of the cost is increasing; as, at 3% premium, the ratio is $\tfrac{103}{100}$; if purchased at a discount the ratio of the cost is diminishing; as, at 3% discount, the ratio is $\tfrac{97}{100}$; hence, if I purchase a draft for $100 at 3% premium, the cost is $100 $\times \tfrac{103}{100} = \$103$; and if I purchase a draft for $100 at 3% discount, the cost is $100 $\times \tfrac{97}{100} = \$97$.

COR. 1.—The cost of a draft, when at a premium, is its face multiplied by an increasing ratio whose denominator is 100 and numerator 100 increased by the premium.

COR. 2.—When there is a discount, the cost of a draft is its face multiplied by a diminishing ratio whose denominator is 100 and numerator 100 diminished by the discount.

REM.—Reckoning by the old par value, £1 is equal $\tfrac{40}{9}$ dollars, and $1 equal £$\tfrac{9}{40}$. The par value of £1 is now fixed by Act of Congress at $4.8665.

PERCENTAGE.

PRACTICAL EXAMPLES.

1. At what time may the balance on the following accounts be paid at one time, without loss to either party?

Dr.	JOHN JONES.	*Cr.*
January 1st, $250	January 15th, $100
February 15th, $300	February 15th, $200
March 1st, $200	March 15th, $150
March 20th, $100	
April 1st, $300	

Ans. In 55 days; that is, February 25th.

2. Average the time in this account.

Dr.	JOHN THOMPSON.	*Cr.*
Oct. 1st, Due $300	Oct. 21st, By cash $300
Oct. 21st, " $500	Dec. 1st, " $400
Nov. 16th, " $400	Dec. 11th, " $200
Dec. 1st, " $600	Jan. 1st, " $50
Dec. 11th, " $200	
Jan. 1st, " $100	

Ans. Due in 34 days from October 1st; that is, November 4th.

PERCENTAGE.

What is the bank discount and also the true discount of the following notes; viz.,

3. Principal $300, due in 3 months, 3 days grace in each case, at 6%?
Ans. Bank discount, $4.65; True discount, $4.58.

4. Principal $500, due in 2 months, 3 days grace in each case, at 6%?
Ans. Bank discount, $5.25; True discount, $5.20.

5. Principal $2500, due in 4 months, 3 days grace in each case, at 6%?
Ans. Bank discount, $51.25; True discount, $50.22.

6. Principal $6000, due in 1 year 3 months, 3 days grace in each case, at 6%?
Ans. Bank discount, $453.00; True discount, $421.20.

7. What is the present worth of a note for $2000, due in 20 months, true discount, at 6%, no grace?
Ans. 1818.18\frac{2}{11}$.

8. How large a note, due in 2 years 6 months, with interest at 6%, will cancel a debt of $3450, due at the same time without interest? *Ans.* $3000.

9. What principal at 6% interest, will cancel a debt of $6000, due in 3 years 4 months? *Ans.* $5000.

10. What would be the bank discount on the debt, at the same rate for the same time without grace?
Ans. $1200.

Rem. 1.—In true discount, the debt is the sum of the principal and interest; thus, $106 due in 1 year at 6%; the principal or present worth is $100, and $6 is the interest on this principal; thus, $100 × $\frac{6}{100}$ = $6, and 100 + 6 = $106; and the fraction by which the debt is to be multiplied has 100 for its numerator and 100 + the interest on $100 for the time and rate for its denominator.

Rem. 2.—Bank discount has for its principal the face of the note given for the debt.

11. What is the time of maturity of a note dated Jan. 10th, 1872, discounted in bank for 90 days?
Ans. April 12th.

12. If the note were discounted for 3 months, what would be the date of its maturity? *Ans.* April 13th.

Rem.—When the time that a note has to run is expressed in days, the note will fall due in the exact number of days and 3 days grace added; but when the time is expressed in months, the note will be due three days later in the month in which it falls, than in that in which it is given?

In Ex. 11, the year 1872 was leap year.

13. A note for $5000, due April 1st, 1872, has the following endorsements: April 1st, 1873, Received $500 on note; April 1st, 1874, Received $1000; Oct. 16th, 1875, Received $1000; April 16th, 1877, Received $1000; how much was due on Oct. 16th, 1878, interest at 6%? *Ans.* $3028.12.

14. A note of $6000, due Jan. 1st, 1871, interest at 6%, is endorsed: July 1st, 1871, Received $100; Jan. 1st, 1872, Received $100; June 19th, 1872, Received $100; Dec. 15th, 1872, Received $100; May 20th, 1873, Received $150; Oct. 30th, 1873, Received $150; Jan. 21st, 1874, Received $200. What was due Jan. 1st, 1875?

$6000 × $\frac{6}{100}$ × 4 = $1440   Interest.
                            900   Payments.
                          ─────
                          $540   Balance on Interest.
                          6000   Principal
                          ─────
                          $6540   Due Jan. 1st, 1875.

PERCENTAGE. 191

COR.—As the payments in Ex. 14 never exceeded the interest; their sum can be deducted from the interest for the whole time. If at any time the sum of the payments exceeded the interest to that time, then the interest to that time must be computed and added to the principal, from which the sum of the payments must be subtracted, and the remainder taken as principal.

REM.—Special attention should be given to the principle of partial payments and the examples given, which will obviate the necessity of many similar examples.

15. How much cash will pay a debt of $1005, of which $475 is to be paid in 10 months, and the remainder in 15 months, interest at 6%, true discount?
Ans. $945.40.

16. Sold 200 bbl. pork, commission $4\frac{1}{2}\%$, freight $100, and remitted $2776.25; what was the pork sold for per barrel? *Ans.* $15.

17. If I fall 10% from my retail price, which is 30% advance on the cost, what per cent will my profit be?
Ans. 17%.

18. Took a risk of $40000 at $\frac{3}{4}\%$; reinsured $\frac{1}{2}$ of it in another company at $\frac{5}{8}\%$, and $\frac{1}{4}$ of it at $\frac{7}{8}\%$; what did I gain by reinsuring? *Ans.* $22\frac{1}{2}$.

19. Bought a farm for $18000, one-third cash and the balance in two equal annual installments, which were discounted in bank without grace at 6%; what amount of cash pays for the farm? *Ans.* $16920.

20. Sold a quantity of goods for $1800, which was 10% less than cost; what would have been the gain per cent if I had sold them for $2400? *Ans.* 20%.

21. What is the cost of a draft for $3000, at ¼% premium, payable in 60 days and 3 days grace, 6% interest allowed? *Ans.* $2991.

REM.—The interest and the premium must be computed separately on the face of the draft.

22. In a section of land (640 acres) 30% is in wheat, 25% pasture, one-fourth meadow, and the balance woodland; how much wood-land was there? *Ans.* 128 acres.

23. The population of a city in 1870 was 240,000, which was 20% more than in 1860; and in 1865 it was 5% more than in 1860; what was the population in 1865?
Ans. 210,000.

24. A cargo of damaged goods was sold for $20,000, which was 33⅓% less than cost; what was the cost?
Ans. $30,000.

25. A merchant sells goods for $26,400, which was at a profit of 20%; what did the goods cost?
Ans. $22,000.

26. What is the commission on $56,000 at 1%? At 2%? 3%? 4%? ¼%? ½%? ¾%?
Ans. $56; $112; $168; $224; $14; $28; $42.

27. What per cent is any number of itself? Of double itself? Of three times itself? Of ½ itself? Of ¼ itself?
Ans. 100%; 50%; 33⅓%; 200%; 400%.

28. If one number is 10% of another number, what per cent is the latter of the former? *Ans.* 1000%.

Thus, 10 is 10% of 100, and 10 is 100% of itself (10), and 100 is 1000% of 10.

29. A merchant made a profit of $200, which was 20% of his investment; what was the investment?
20% = $200; 100% = $1000. *Ans.* $1000.

PERCENTAGE.

30. Six men purchased 640 acres of land; the first is to have 6⅜%, the second 8⅓%, the third 10%, the fourth 12½%, the fifth 16⅔%, the sixth the balance; how many acres? *Ans.* 293¼ acres.

31. Sold a lot of grain, and after deducting commissions of 4½% for selling and 2½% for buying, I invested the proceeds $7640 in sugar; what was the grain sold for? *Ans.* $8200.

$$\$7640 \times \frac{102\tfrac{1}{2}}{100} \times \frac{100}{95\tfrac{1}{2}}$$

32. Sold goods 40% off the marked price, and thereby lost $1200, which was 10% of cost; what was the marked price and cost?
Ans. Marked price $18000; cost $12000.

33. What is the cost of a draft on New York for $2000, at ¾% premium? *Ans.* $2015.

34. What is the cost of a draft on New Orleans for $2500, at ½% discount? *Ans.* $2487½.

35. A merchant entered business with $5000; the first year he gained 20%, which he added to his capital; the second year he gained 20%, and added it to his capital; the third year he lost 25%; what had he then?
Ans. $5400.

36. Sold goods for $375, thereby losing $25; what per cent did I lose? *Ans.* 6¼%.

37. After selling 25 sheep there remained 225; what per cent were sold? *Ans.* 10%.

38. A tax collector gets 5% additional on the taxes which he collects; the cost on property valued at $56000 was $294; what was the rate per cent of the tax?
Ans. ½%.

39. In a storm 20% of the cargo was thrown overboard; at what rate of gain must the remaining cargo be disposed of so that no loss is sustained? *Ans.* 25%.

40. A merchant commenced business with $3000 and closed with $4500; what per cent did he gain?

Ans. 50%.

41. What per cent of £3 10s. is 14s.? *Ans.* 20%.
42. What per cent of 5 lb. Avoirdupois is 8 oz.?

Ans. 10%.

43. What per cent of 5 lb. Avoirdupois is 14 pwt. 14 gr.? *Ans.* 1%.
44. What per cent of 1 ton is 1 cwt.? *Ans.* 5%.
45. One quart is what per cent of 1 bushel?

Ans. $3\tfrac{1}{8}$%.

46. What is the cost of a draft on New Orleans for $2000, at sixty days, with grace, discount $\tfrac{1}{2}$%, and interest allowed at 6%? *Ans.* $1969.

47. 2 is what per cent of 5? *Ans.* 40%.
48. $3\tfrac{1}{2}$ is what per cent of $5\tfrac{1}{2}$?

$$\frac{7}{11} = \frac{63\tfrac{7}{11}}{100} = 63\tfrac{7}{11}\%, \; Ans.$$

49. $13\tfrac{1}{2}$ is what per cent of $9\tfrac{1}{2}$?

$$\frac{27}{19} = \frac{142\tfrac{2}{19}}{100} = 142\tfrac{2}{19}\%, \; Ans.$$

REM.—A common fraction is reduced to hundredths by multiplying both terms of the fraction by the quotient obtained by dividing 100 by the denominator of the fraction.

A LLIGATION.

Alligation is the mixing of different qualities of grain, groceries, liquors, etc., in order to get an article of a certain price; thus, sugar at 5 cts. and 9 cts. per lb. may be mixed together in such proportions as to make an article of any value between the two given prices.

Cor.—The mixture cannot be made of less value than 5 cts. nor more than 9 cts.; for if a quantity be taken at 5 cts., and some of 9 cts. be added, it will increase the value above 5 cts.; and if 5 ct. sugar be added to the 9 ct., it will diminish the value.

Rem.—If both kinds are either above or below the required price, the mixture cannot be made

PROBLEM.

1. What relative quantity of each must be taken of two kinds of sugar, the one worth 5 cts. per lb., and the other 11 cts., in order that the mixture be worth 7 cts. per lb.?

$$7 \begin{Bmatrix} 5 \\ 11 \end{Bmatrix} \begin{matrix} \frac{1}{2} \\ \frac{1}{4} \end{matrix} \begin{matrix} 2 = 10 \\ \underline{1 = 11} \\ 3 \div 21 \\ 7 \text{ cts.} \end{matrix}$$

ALLIGATION.

EXEMPLIFICATION.—For every lb. at 5 cts. there is a gain of 2 cts., and for $\frac{1}{2}$ lb. a gain of 1 ct.; for every lb. at 11 cts. there is a loss of 4 cts., and of $\frac{1}{4}$ lb. the loss is 1 ct.; hence, if $\frac{1}{2}$ lb. at 5 cts. be taken, and $\frac{1}{4}$ lb. at 11 cts., the gain and loss will be equal; this is then the ratio; or, reducing the fractions to a common denominator and canceling the denominators, the ratio is 2 at 5 to 1 at 11.

REM.—It matters not how many different qualities are to be mixed, only two at a time can be mixed; ∴ the principle developed in the above problem is the only principle in alligation.

2. Mix together coffee worth 17 cts., 19 cts., 21 cts., and 24 cts., so that the mixture shall be worth 20 cts.

$$20\begin{cases} 17 \\ 21 \end{cases} \Big| \begin{matrix} \frac{1}{2} \\ 1 \end{matrix} \Big| \begin{matrix} 1 \\ 3 \end{matrix} \Big| \begin{matrix} = 17 \\ = 63 \end{matrix}$$
$$\phantom{20\{17\}}4 = 80$$
$$\phantom{20\{17\}}1 = 20$$

$$20\begin{cases} 19 \\ 24 \end{cases} \Big| \begin{matrix} 1 \\ \frac{1}{4} \end{matrix} \Big| \begin{matrix} 4 \\ 1 \end{matrix} \Big| \begin{matrix} = 76 \\ = 24 \end{matrix}$$
$$\phantom{20\{19\}}5 = 100$$
$$\phantom{20\{19\}}1 = 20$$

$$20\begin{cases} 17 \\ 24 \end{cases} \Big| \begin{matrix} \frac{1}{3} \\ \frac{1}{4} \end{matrix} \Big| \begin{matrix} 4 \\ 3 \end{matrix} \Big| \begin{matrix} = 68 \\ = 72 \end{matrix}$$
$$\phantom{20\{17\}}7 = 140$$
$$\phantom{20\{17\}}1 = 20$$

$$20\begin{cases} 19 \\ 21 \end{cases} \Big| \begin{matrix} 1 \\ 1 \end{matrix} \Big| \begin{matrix} = 19 \\ = 21 \end{matrix}$$
$$\phantom{20\{19\}}2 = 40$$
$$\phantom{20\{19\}}1 = 20$$

It is evident that 17 and 19 cannot make a mixture worth as much as 20 cents; it is also evident that 21 and 24 cannot make a mixture worth as little as 20 cts.; but 17 and 21, 19 and 24, 17 and 24, and 19 and 21 may be mixed so as to be worth 20 cts.

By the first is taken 1 at 17 to 3 at 21; and 4 at 19 to 1 at 24; again, 4 at 17 to 3 at 24; and 1 at 19 to 1 at 21.

All the mixtures being of the same value, may be put together, and the whole will be worth 20 cts.; thus,

ALLIGATION. 197

$$
\begin{array}{c|ccc}
17 & 1+4=5 & = & 85 \\
21 & 3+1=4 & = & 84 \\
19 & 4+1=5 & = & 95 \\
24 & 1+3=\underline{4} & = & \underline{96} \\
& 18 & = & 360 \\
& 1 & = & 20
\end{array}
$$

The result is the same; that is, the whole mixture is worth 20 cts.

Take the same articles to form a mixture, and place them together thus:

$$
20 \left\{ \begin{array}{l} 17 \cdots \\ 19 \cdots \\ 21 \cdots \\ 24 \cdots \end{array} \right.
\begin{array}{|c|c|} \frac{1}{3} & 1 \\ 1 & 4 \\ 1 & 3 \\ \frac{1}{4} & 1 \end{array}
\text{ The same as before.}
$$

$$
20 \left\{ \begin{array}{l} 17 \cdots \\ 19 \cdots \\ 21 \cdots \\ 24 \cdots \end{array} \right.
\begin{array}{|c|c|} \frac{1}{3} & 4 \\ 1 & 1 \\ 1 & 1 \\ \frac{1}{4} & 3 \end{array}
\text{ The same.}
$$

Cor.—The same relative portions of each is obtained by taking the difference between each price and the mean price, and placing this difference opposite the price to which each one is connected. Solve the next one according to this corollary.

8. A vintner has four qualities of wine, viz., at $1.30, $1.50, $1.75, and $1.95 per gallon; he has an order for wine at $1.60; what relative quantity of each must be put into the mixture?

198   ALLIGATION.

$$160 \begin{cases} 130 \cdots & 15 \\ 150 \cdots & 35 \\ 175 \cdots & 30 \\ 195 \cdots & 10 \end{cases} \qquad 160 \begin{cases} 130 & \frac{1}{30} & 15 \\ 150 & \frac{1}{10} & 35 \\ 175 & \frac{1}{15} & 30 \\ 195 & \frac{1}{30} & 10 \end{cases}$$

The results are the same; but observe that the fraction must not be reduced to the least common denominator, but only to their common denominator, the one multiplied by the other; of course, the mixtures would be of the same value, and each pair would have the same relative value, but the four would not correspond.

REM.—Observe that different results may be obtained by different methods, and all will be correct.

Sometimes a merchant wishes to put the whole of a certain kind into the mixture; in which case this article may be mixed with every kind that is on the opposite side of the mean.

4. A merchant has five qualities of liquors, at the following prices per gallon, viz., $1.25, $1.45, $1.60, $1.80, and $1.90, and he has an order for liquor at $1.50 per gallon. Of the liquor at $1.90 he has 40 gallons, all of which he wishes to put into the mixture; how much must be taken of each of the other kinds?

$$150 \begin{cases} \cdots 125 \cdots & 10 + 40 \\ 145 \cdots & 30 \\ 160 \cdots & 25 \\ 180 \cdots & 5 \\ 190 & 25 \end{cases} \quad \begin{array}{rcl} 50 \times \frac{8}{5} &=& 80 = 100.00 \\ 30 \times \frac{8}{5} &=& 48 = 69.60 \\ 25 \times \frac{8}{5} &=& 40 = 64.00 \\ 5 \times \frac{8}{5} &=& 8 = 14.40 \\ 25 \times \frac{8}{5} &=& \underline{40} = \underline{76.00} \\ & & 216 = 324.00 \\ & & 1 = 1.50 \end{array}$$

$$25 \times \tfrac{40}{125} = \tfrac{8}{5}.$$

ALLIGATION. 199

5. A grocer has spices at 18, 24, 36, and 42 cts. per lb., of which he wishes to make a mixture worth 32 cts.

$$32 \begin{cases} 18 \cdots & 10 \\ 24 \cdots & 4 \\ 36 \cdots & 8 \\ 42 \cdots & 14 \end{cases}$$

COR. 1.—Connect any two rates, one of which is less and the other greater than the mean. Take the difference of each of these and the mean, and place the difference opposite the price of the one with which it is connected.

COR. 2.—When there are odd terms, that is, when those above and below the mean price are not of equal numbers, the odd term will be connected with an opposite one that is already connected, and instead of having only one difference opposite to it, will have two or more; these several numbers must be added together, and their sum will be the quantity to be taken of the price to which it stands opposite.

6. How many pounds of each kind of tea, of the values of 30 cts., 35 cts., 40 cts., 52 cts., 55 cts., and 60 cts., must be taken to make a mixture worth 45 cts.?

$$45 \begin{cases} 30 \cdots & 7 \\ 35 \cdots & 10 \\ 40 \cdots & 15 \\ 52 \cdots & 15 \\ 55 \cdots & 10 \\ 60 \cdots & 5 \end{cases}$$

REM.—The results will be correct, although the portions will be different by making other connections.

INVOLUTION AND EVOLUTION

Involution and *Evolution* correspond very nearly to Multiplication and Division. Involution consists in the multiplication of the same number by itself, or of the same factor entering two or more times into a product; whilst Evolution takes the product and restores the equal factors of which it is composed.

Involution is called the raising of powers; thus,

$2 \times 2 = 4$ is called the 2d power of 2.
$2 \times 2 \times 2 = 8$ is called the 3d power of 2.
$3 \times 3 = 9$ is called the 2d power of 3.
$3 \times 3 \times 3 = 27$ is called the 3d power of 3.

Evolution, or the extracting of roots, is exactly the reverse of Involution.

In order to extract the second or square root, we take the second power and from it restore the factors; thus, 4 is the second power of 2; hence the square root of 4 is 2; and the square root of 9 is 3; of 16 it is 4; of 25 it is 5; of 36, 6; of 49, 7; of 64, 8, etc.

The number given is a square surface, each side of which is a factor; the sides of the square being equal, the factors are equal; thus,

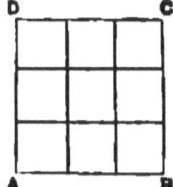

The figure ABCD is a square surface, 3 inches in length and 3 inches in breadth; the angles at A, B, C, and D are equal, that is, from each of these points the sides have the same divergence; hence, if any one be placed on the other, the two will coincide.

AB is 3 inches long, and for every inch in breadth there are 3 square inches, and since AD and BC are each 3 inches, there will be 3 times 3 sq. in. = 9 sq. in.

In the square root we have given the area of the square surface to find a side; and as $3 \times 3 = 9$, we know that 9 is composed of the two equal factors of 3. ∴ 3 is the square root of 9.

There are many numbers of which we cannot obtain an exact root, as we can only extract the root exactly of 4, 9, 16, 25, 36, 49, 64, 81, 100, in 100; that is, 9 have exact roots and 91 have not. Of these we can approximate decimally, or we can put them under the radical sign, thus, $\sqrt{3}$, $\sqrt{5}$, $\sqrt{6}$, by which the root is expressed, and they are read, the square root of 3, or 5, or of 6, as the case may be.

In larger numbers, it is more difficult to get the root; thus, $11 \times 11 = 121$; $12 \times 12 = 144$; $13 \times 13 = 169$, etc. For the extraction of these roots a method will be given hereafter. Sometimes, however, the number is not a perfect square, and hence only an approximate root can be found.

EVOLUTION.

2	3	4	5	6	7	8
4	9	16	25	36	49	64
9	10	11	12	13	14	15
81	100	121	144	169	196	225

The square of the highest digit has two places of figures, the square of the least one; the square of the least number of tens is hundreds, and when the tens reach 4 it has four places of figures, never more.

```
   1        11         9         99
   1        11         9         99
  ---      ----       ---      -----
   1       121        81       9801
```

An increase of one figure in the root makes an increase of two in the square, for the square of the least unit is a unit, and of the least units and tens is hundreds; the square of the largest digit is tens, and the square of the largest tens is thousands.

Therefore, if the square be pointed off in periods of two figures each, there will be as many periods as figures in the root.

```
   1,21 ( 10 + 1           10 × 10 = 100
   1 00                    10 ×  2 =  20
  -------                   1 ×  1 =   1
  21 ) 21                              ---
       21                              121


       10 × 10 = 100         1,44 ( 10 + 2
       10 ×  4 =  40         1 00
        2 ×  2 =   4        -------
                 ---        22 ) 44
                 144              44
```

EVOLUTION.

$a = 10$, then $12 = a + b$ $\quad = \quad 10 + 2$
$b = 2,$ $\quad\quad\quad \dfrac{12}{144} \quad \dfrac{a + b}{a^2 + ab}$ $\quad \dfrac{10 + 2}{100 + 20}$

$$\quad\quad\quad + ab + b^2 \quad\quad\quad 20 + 4$$
$$a^2 + 2ab + b^2 (a + b \quad 100 + 40 + 4 (10 + 2$$
$$\underline{a^2} \quad\quad\quad\quad\quad \underline{100}$$
$$2a + b)\, 2ab + b^2 \quad\quad 20 + 2\,)\,40 + 4$$
$$\underline{2ab + b^2} \quad\quad\quad\quad \underline{40 + 4}$$

As in the multiplication the first term is multiplied into itself but once, so one of its equal factors multiplied by itself occupies the first square; the second number is twice multiplied into the first and once into itself; therefore the divisor must be twice the first plus the second. This is observable in the two rectangles, each of which has a for its length and b for its breadth; and one side of the little square (b^2) is b, and $2a + b$ make the full length of the rectangle, which together with the square (a^2) make the whole of the large square $(a+b)^2$, and its breadth is b, which is the term wanting in the root.

$$\text{Let } a + b + c \quad = \quad 100 + 20 + 3$$
$$\dfrac{a + b + c}{a^2 + ab + ac}$$
$$ab + b^2 + bc$$
$$\underline{ac + bc + c^2}$$
$$a^2 + 2ab + b^2 + 2ac + 2bc + c^2\,(\,a + b + c$$
$$\underline{a^2}$$
$$2a + b\,)\; + 2ab + b^2$$
$$\underline{2ab + b^2}$$
$$2a + 2b + c\;)\;2ac + 2bc + c^2$$
$$\underline{2ac + 2bc + c^2}$$

$$100+20+3$$
$$100+20+3$$
$$\overline{10000+2000+300}$$
$$2000+400+60$$
$$300+60+9$$
$$\overline{10000+4000+400+600+120+9}$$

$$\begin{array}{r}123\\123\\\hline 369\\246\\123\\\hline\end{array}$$

$1,51,29\ (100+20+3$
$\underline{1\ 00\ 00}$

$200\ )\ \overline{51\ 29}$
$\underline{20}\quad 44\ 00$
$\overline{220}$

$240\ )\ \overline{729}$
$\underline{\ \ 3}\quad 729$
$\overline{243}$

Cor.—In order to extract the square root of any number, point it off in periods of two figures each, beginning with units, the left hand period may have but one figure ; find the greatest figure whose square will either be equal or less than the number in the left-hand period regarded as units, place this figure in the root and subtract its square from the left-hand period, and to the remainder, if there is any, annex the next period ; if there is no remainder, the next period itself will be the dividend ; and for a trial divisor, double the root already found, regarding it as tens, as we are now considering it as with a figure annexed to which it will hold the tens' place ; find how often the trial divisor is contained in the dividend and annex that figure to the root already found, and add this last figure of the root to the trial divisor, which makes the divisor complete ; perform the division and if there are other periods bring them down in like manner, and repeat the above method until the whole root is found.

EXAMPLES.

1. Extract the square root of 9801.
2. Extract the square root of 103041.
3. Extract the square root of 197136.
4. Extract the square root of 998001.
5. Extract the square root of 603729.

As a fraction is squared by multiplying it by itself, thus, $\frac{4}{5} \times \frac{4}{5} = \frac{16}{25}$, so its root is extracted by extracting the roots of both its terms; hence, $\sqrt{\frac{16}{25}} = \frac{4}{5}$, $\sqrt{\frac{25}{36}} = \frac{5}{6}$, $\sqrt{\frac{36}{49}} = \frac{6}{7}$, etc.

By a proposition in Geometry, it is proved that the square described on the hypothenuse of a right-angled triangle is equivalent to the sum of the squares described upon the base and perpendicular; thus,

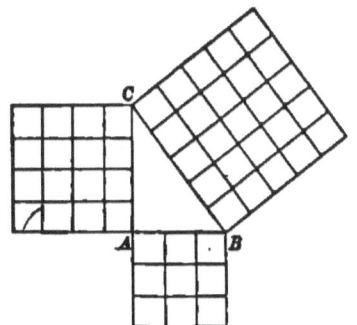

Let ABC be a right-angled triangle, AB the base, AC the perpendicular, and BC the hypothenuse. AB is 3 feet long, and contains 9 square feet; AC is 4 ft., and contains 16 sq. ft.; and BC is 5 ft., and contains 25 sq. ft., equal to the sum of $\overline{AB}^2$ and $\overline{AC}^2$.

This figure is exemplified by the walls of a house, which are always perpendicular to the surface of the earth or to the street. If the foot of a ladder rest on the ground some distance from a house, and the top of the ladder against the house, the distance of the foot of the ladder from the house is the base, the height of the house from the ground to the top of the ladder is the perpendicular, and the ladder is the hypothenuse.

EXAMPLES.

1. A ladder 25 feet long, whose foot is 15 feet distant from the house, just reaches the top of the house. How high is the house?

Let b = base.
p = perpendicular.
h = hypothenuse.

$$b^2 + p^2 = h^2$$
$$p^2 = h^2 - b^2 \therefore 25 \times 25 = 625$$
$$p = \sqrt{h^2 - b^2} \quad 15 \times 15 = 225$$
$$b = \sqrt{h^2 - p^2} \qquad \overline{4{,}00} \; (20 = p.$$
$$\phantom{b = \sqrt{h^2 - p^2} \qquad} 4$$
$$\phantom{b = \sqrt{h^2 - p^2} \qquad} \overline{)\,00}$$

2. What is the length of the diagonal of a square, each side of which is 12 feet?

The diagonal of a square is the same as the hypothenuse, having base and perpendicular the same.

$$12 \times 12 = 144$$
$$12 \times 12 = 144$$
$$\sqrt{288} =$$

3. What is the length of the diagonal of a rectangle whose sides are 45 and 60?

EVOLUTION.

4. A ladder 75 feet long being placed with its foot in the street reaches a window on one side 45 feet high, and on the other side 60 feet high. How wide is the street? *Ans.* Street, 105.

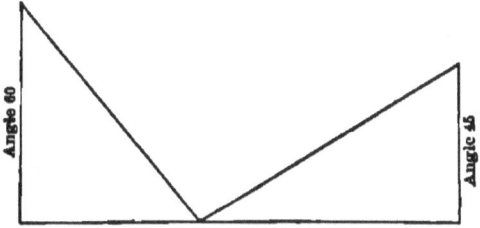

When a number is both integral and decimal, as 455.742, the integers must be pointed off as if there were no decimals, and the decimals as if there were no integers; thus, 4'55.74'20'. Add more ciphers if necessary.

To extract the square root of a fraction whose denominator is not a perfect square, multiply both terms of the fraction by the denominator; thus, to extract the root of $\frac{3}{5}$. $\frac{3}{5} \times 5 = \frac{15}{25}$.

The root of the denominator is now 5; of the numerator,

```
        15.0000 ( 3.873
         9
   68 )  600
         44
  767 ) 5600
        5369
  774 ) 23100
```

The result = $\frac{3.873}{5}$ = .7746 nearly.

CUBE ROOT.

The *Cube* is a solid having for its base a square and every side equal to the base all equal squares; it has therefore three dimensions, all equal; and as 1 foot is 12

inches, let the base be a square, each side 12 inches; the square will contain 144 square inches, and for every inch in height 144 solid inches, and 12 in height 1728 solid inches; that is
$$12 \times 12 \times 12 = 1728.$$
The cube of 1 is 1, and of 2 it is 8.
$$1 \times 1 \times 1 = 1, \text{ or } 2 \times 2 \times 2 = 8,$$
and the cube of 12 is 1728 increased three figures.

$9 \times 9 \times 9 = 729$, and $99 \times 99 \times 99 = 970299$. The cube of the largest digit has three places of figures, and the cube of two places of the largest digits has six places of figures; therefore, the increase for one figure in the root is three in the cube; hence the pointing off in periods of three figures.

Extracting the cube root consists in having given the cube or solid contents to find a side.

REM.—As the cube is the product of three dimensions, and as the root is one of those dimensions, the divisor must necessarily have two dimensions.

As $11 \times 11 \times 11 = 1331 = (10 + 1)^3 = 1331.$
Literal $a = 10, b = 1.$

$$
\begin{array}{r}
a + b \\
a + b \\ \hline
a^2 + ab \\
 + ab + b^2 \\ \hline
a^2 + 2ab + b^2 \\
a + b \\ \hline
a^3 + 2a^2b + ab^2 \\
 a^2b + 2ab^2 + b^3 \\ \hline
\end{array}
$$

$a^3 + 3a^2b + 3ab^2 + b^3\ (\ a+b$
a^3

$3a^2 + 3ab + b^2\ )\ 3a^2b + 3ab^2 + b^3$
$ 3a^2b + 3ab^2 + b^3$

EVOLUTION.

Cor.—The first term of the root is a, the cube of which is a^3; the second term of the root is b, and the required divisor must be contained b times in what remains after deducting a^3; therefore it must be $3a^2 + 3ab + b^2$; that is, 3 times the square of the first term of the root, three times the product of the last term and the previous root, plus the square of the last term of the root; by extending the exemplification it is only necessary to repeat the foregoing, as each successive divisor is formed in the same way.

PROBLEMS.

1. Extract the cube root of 1331, which power is thus found, $11 \times 11 \times 11$. Point off in periods of three figures; the root of the left-hand period is 1; cube it and subtract; *the value* of this 1 is 10, when a figure is annexed; hence,

$3a^2 = 10 \times 10 \times 3 = 300$ 1,331 (11
$3ab = 10 \times 1 \times 3 = 30$ 1
$b^2 = 1 \times 1 = \underline{1}$ $\overline{331}$
$ \underline{331}$ $\underline{331}$

2. Extract the root of 970299; point in periods. The root of the first period is 9; cube it and subtract, then bring down the next period and form a divisor; thus,

$3a^2 = 90 \times 90 \times 3 = 24300$ 970,299 (99
$3ab = 90 \times 9 \times 3 = 2430$ 729
$b^2 = 9 \times 9 = \underline{81}$ $\overline{241\,299}$
$ \underline{26811}$ $\underline{241\,299}$

Rem.—When there is one figure in the root and a second figure is taken into consideration, then the first figure must be regarded

210 EVOLUTION.

as tens, and a zero put after it; this must be done to the second when a third is considered.

Cor.—In order to extract the cube root of a number, point off the number in periods of three figures each, beginning with units; the left-hand period will have one, two, or three, depending on the number of figures. Next, place in the root the largest figure whose cube is either equal to or less than the number in the left-hand period regarded as units; cube the root thus obtained and subtract its cube from the left-hand period, and to the remainder, if there is any, annex the next period; if there is no remainder, then the next period will be the dividend, and for a trial divisor take three times the square of the root already found regarded as tens; find how often it is likely to be contained in the dividend, after making an allowance for what is to be added to the divisor; annex this figure to the root already found, and add to the trial divisor three times the product of the last figure and the previous root regarded as tens, and to this sum add the square of the last figure of the root; this renders the divisor complete; perform the division, and if there are other periods, bring them down in order and repeat the foregoing process until the entire root is found.

EXAMPLES.

1. Extract the cube root of 1367631.   *Ans.* 111.
2. Extract the cube root of 997002999.   *Ans.* 999.
3. Extract the cube root of 91125.   *Ans.* 45.
4. Extract the cube root of 9.261.   *Ans.* 2.1.
5. Extract the cube root of $\frac{1}{27}$.   *Ans.* $\frac{1}{3}$.

EVOLUTION. 211

6. Extract the cube root of $\frac{8}{125}$.   *Ans.* $\frac{2}{5}$.
7. Extract the cube root of 2515456.   *Ans.* 136.
8. Extract the cube root of 1124864.   *Ans.* 104.
9. Extract the cube root of 8024024008.

                                                   Ans. 2002.

10. Extract the cube root of 65939.264.   *Ans.* 40.4.
11. Extract the cube root of .000343.   *Ans.* .07.
12. Extract the cube root of $5\frac{23}{64}$.   *Ans.* $1\frac{3}{4}$.
13. Extract the cube root of $\frac{5}{9}$.

$$\frac{5}{9} = \frac{15}{27} = \frac{\sqrt[3]{15}}{3} = \frac{2.466+}{3} = .822+, \textit{ Ans.}$$

Rem.—1. The volumes of cubical figures are to each other as the cubes of their edges.

2. In order to extract the cube root of a fraction, first render the denominator a perfect cube.

PRACTICAL EXAMPLES.

1. What is the depth of a measure of 1 bushel U. S. dry measure, in the form of a cube? Of ½ bushel?

Ans. Of a bushel 12.907+ in., of ½ bushel 10.244+ in.

2. What would be the length of a cubical pile of stone equal in volume to a rectangular pile whose length is 64 feet, breadth 27 feet, and 8 feet high?

                                                     Ans. 24 feet.

3. If a ball 4 inches in diameter weighs 10 lb., what is the diameter of a ball weighing 640 lb.?

                                                 Ans. 16 inches.

Rem.—The volume of spheres are to each other as the cubes of their diameters.

4. What are the dimensions of a cube containing 1728 cu. in.? *Ans.* 12 in.

It is demonstrated in Geometry that similar cubical bodies are to each other as the cubes of their like dimensions; as, if a cube of 1 inch weigh 2 lbs., one of three inches would weigh $27 \times 2 = 54$ lbs.

5. How many cubes whose edges measure $\frac{1}{4}$ inch, would be contained in a cubical block whose edges are 2 inches?

$$\tfrac{1}{4} \times \tfrac{1}{4} \times \tfrac{1}{4} = \tfrac{1}{64}.$$
$$2 \times 2 \times 2 = 8 \times \tfrac{64}{1} = 512, \textit{Ans.}$$

6. How many shot, $\frac{1}{16}$ inch in diameter, can be made of a globe of lead 4 inches in diameter?

$$\tfrac{1}{16} \times \tfrac{1}{16} \times \tfrac{1}{16} = \tfrac{1}{4096}.$$
$$4 \times 4 \times 4 = 64 \times \tfrac{4096}{1} = 252144, \textit{Ans.}$$

7. How many cubes whose edges measure $\frac{1}{3}$ in. is contained in a cubical block whose edges are 3 inches?
Ans. 729 cubes.

8. How many square feet in the surface of a cube whose volume is 3375 cubic feet? *Ans.* 1350 sq. ft.

9. What is the length of an edge of a cubical bin which contains 500 bushels of wheat?
Ans. 8 feet 6 inches.

SERIES OF COMMON DIFFERENCES.

A ***Series of Common Differences***, usually termed Arithmetical Progression, is a series in which the difference of any two consecutive terms, taken in order, is the same; as, in the series

  (1.) 1, 2, 3, 4, 5, 6, etc.

The first term is 1, and the common difference is 1. This is an increasing series, as each successive term is larger than the previous one.

       1 2 3 4
  (2.) 1, 3, 5, 7, etc.,

is an increasing series, and the common difference is 2.

       1 2 3 4
  (3.) 2, 5, 8, 11, etc.,

is an increasing series, and the common difference is 3.

  (4.) 12, 10, 8, 6, 4, 2, etc.,

is a decreasing series, with a common difference of 2.

 Rem.—In an increasing series, the common difference is found by subtracting any term, except the last, from the following term. In a decreasing term it is found by subtracting any term, except the first, from the preceding term.

SERIES.

PROBLEM.

The sum of a series of equal differences is equal to one-half the product of the number of terms and the sum of the first and last terms.

As, the sum of ten terms of the series, 1, 2, 3, etc., is

$$\begin{array}{cccccccccc}
1, & 2, & 3, & 4, & 5, & 6, & 7, & 8, & 9, & 10. \\
10, & 9, & 8, & 7, & 6, & 5, & 4, & 3, & 2, & 1. \\
\hline
11, & 11, & 11, & 11, & 11, & 11, & 11, & 11, & 11, & 11.
\end{array}$$

Sum of all the terms of both series $= 11 \times 10 = 110$, and sum of 1 series,

$$\frac{11 \times 10}{2} = 55.$$

The series may also be written,

$$\overset{1}{a},\ \overset{2}{a+d},\ \overset{3}{a+2d},\ \overset{4}{a+3d},\ \text{etc.,}$$

or, $\quad a,\ a-d,\ a-2d,\ a-3d,\ \text{etc.}$

In these series a represents the first term and d the common difference.

Cor.—Any term in an increasing series is equal to the first term plus the difference taken once less than the number of terms, and in a decreasing series any term is equal to the first term minus the common difference taken as often as the number of the term minus 1, and any term may be regarded as the last term. Therefore, the expression for the last term in the former is

$$l = a + (n-1)d.$$

SERIES.

The expression for the last term in the latter is

$$l = a - (n - 1) d.$$

The last term is equal to the first plus or minus the product of the difference and the number of terms less one.

EXAMPLES.

1. Find the ninth term in the series 1, 3, 5, 7, etc.

FORMULA.—$l = a + (n - 1) d = 1 + (8 \times 2) = 17 =$ ninth term.

2. Find the tenth term of the series 2, 4, 6, 8, etc.
Ans. 20.

3. Find the the fifth term of 12, 10, 8, 6, etc.

$$l = 12 - 8 = 4, \textit{ Ans.}$$

4. Find the sum of six terms of the series 1, 3, 5, 7, etc.
$$l = 11;$$
$$S = \tfrac{6}{2} (1 + 11) = 36.$$

5. Find the sum of seven terms of 15, 13, 11, etc.

$$l = a - (n - 1) d = 15 - 12 = 3.$$
$$S = \tfrac{7}{2} (15 + 3) = 63.$$

6. Find the sum of 100 terms of the series 1, 3, 5, 7, 9, etc. $\quad S = 10000.$

7. Find the sum of a decreasing series, whose first term is 12 and the common difference $\tfrac{1}{4}$. $\quad S = 150.$

8. Insert three terms of a series between $\tfrac{1}{6}$ and $\tfrac{1}{4}$.
$$\tfrac{3}{16}, \tfrac{5}{18}, \tfrac{11}{24}.$$

216 *SERIES.*

9. The first term of a series is 1 and the number of terms 23; what must be the common difference in order that the sum may be 149½? Com. diff. = ½.

FORMULA 1. $S = \dfrac{n}{2}(a + l)$.

$$149\tfrac{1}{2} = \tfrac{23}{2} + \tfrac{23}{2}l;$$
$$299 = 23 + 23l;$$
$$276 = 23l;$$
$$12 = l.$$

FORMULA 2. $l = a + (n - 1)d$.

Subtract a from both members of Formula 2 and divide by $(n - 1)$,

$$l - a = (n - 1)d.$$
$$d = \dfrac{l - a}{n - 1} = \tfrac{11}{22} = \tfrac{1}{2}.\ \ Ans.$$

The common difference is equal to the last term minus the first divided by the number of terms less one.

REM.—When terms are to be inserted, find the common difference and then apply it.

10. What is the sum of an increasing series whose first term is ½, the common difference ½, and the number of the terms 20? Sum = 105.

11. What is the sum of an increasing series whose first term is 2, common difference 3, and number of terms 12?
$$S = 222.$$

SERIES OF EQUAL RATIOS.

A *Series of Equal Ratios*, improperly called Geometrical Progression, is one in which the ratios of any two consecutive terms taken in the same order is equal; as, 1, 2, 4, 8, 16, is an increasing series, and any term divided by the one immediately preceding it gives a quotient of 2.

Observe the series, $\overset{1}{1}, \quad \overset{2}{2}, \quad \overset{3}{4}, \quad \overset{4}{8}, \quad \overset{5}{16}, \quad \overset{6}{32}.$
Write it literally, $\quad a, \quad ar, \quad ar^2, \quad ar^3, \quad ar^4, \quad ar^5.$
$a = $ 1st term, and $r = $ ratio.

Each term is equal to the product of the 1st term and the ratio raised to the power of the number of terms less one. FORMULA (1). n^{th} term $= ar^{n-1}$.

PROBLEM.

To find the sum of a series of equal ratios.

Take the literal series and form an equation,
$$S = a + ar + ar^2 + ar^3 + ar^4 + ar^5, \qquad (1)$$
and multiply both terms by r,
$$Sr = ar + ar^2 + ar^3 + ar^4 + ar^5 + ar^6. \qquad (2)$$
Subtract the 1st equation from the 2d,
$$Sr - S = ar^6 - a = a(r^6 - 1).$$
Factor the terms,
$$(r - 1)S = a(r^6 - 1).$$

Divide both members by $r - 1$,

FORMULA (2), $S = \dfrac{a(r^s - 1)}{r - 1}.$

In this formula the ratio is raised to the power of the number of terms, which may be designated by n, and the formula will become,

$$S = \dfrac{a(r^n - 1)}{r - 1}. \qquad (3)$$

In this formula, substitute the formula of last term; thus,

n^{th} term or $l = ar^{n-1}.$

Multiply both members by r,

$rl = ar^n.$

$$S = \dfrac{a(r^n - 1)}{r - 1} = \dfrac{ar^n - a}{r - 1} = \dfrac{rl - a}{r - 1}. \qquad (4)$$

COR.—The sum of the series is equal to the product of the ratio and last term, diminished by the first term and divided by the ratio less one.

EXAMPLES.

1. Find the 10th term in the series, 1, 2, 4, 8, 16, etc.

FORMULA. n^{th} term $= ar^{n-1} = 1 \times 2^9$
$= 1 \times 2 \times 2 \times 2 \times 2 \times 2 \times 2 \times 2 \times 2 \times 2 = 512.$

2. Find the 6th term of 64, 32, 16, etc. The ratio is $\frac{1}{2}$.

n^{th} term $= ar^{n-1}.$

6th $= 64 \times (\frac{1}{2})^5 = 64 \times \frac{1}{2} \times \frac{1}{2} \times \frac{1}{2} \times \frac{1}{2} \times \frac{1}{2} = 2.$

3. Find the sum of 5 terms of 1, 2, 4, 8, etc.

$$S = \dfrac{a(r^n - 1)}{r - 1} = \dfrac{1(2^5 - 1)}{2 - 1} = \dfrac{1(31)}{1} = 31.$$

4. Find the sum of 8 terms of 1, 3, 9, 27, etc.

COMPOUND INTEREST.

Compound Interest consists in adding the interest to the principal as often as the interest becomes due, until the end of the time at which it is at interest.

When a sum of money is at interest for a considerable time, and no interest has been collected, although it was specified that the interest was payable annually, semi-annually, or quarterly, and the interest is computed on the principal for the specified time of interest, and at the end of such period added to the principal, this is called Compound Interest; thus, the interest of $1 at 6% for 1 year is $.06, which if added to the principal makes $1.06. The ratio will be $\frac{106}{100}$, or 1.06; therefore,

$1.06 is the amt. of $1 for 1 yr. at 6% $= R$.
1.06
―――――
1.1236   "   "   2 yrs. at 6% $= R^2$.
1.06
―――――
1.191016   "   "   3 yrs. at 6% $= R^3$.
1.06
―――――
1.26247696   "   "   4 yrs. at 6% $= R^4$.
1.06
―――――
1.3382255776   "   "   5 yrs. at 6% $= R^5$.
1.06
―――――
1.418519112256   "   "   6 yrs. at 6% $= R^6$.
1
―――――
.418519, called 42 cents, is the compound interest of $1 for 6 years at 6%.

COMPOUND INTEREST.

Let $R = 1.06$, then R^2, etc. will be the amt. of $1 for one, two, etc. years, R^n for n years.

If the interest were payable semi-annually, it would be 1.03 to the power expressed by double the number of years; thus,

$$1.03 = R.$$

$$(1.03)^{12} = R^{12}.$$

If quarterly, $(1.015)^{24} = R^{24}$.

R representing respectively 1.06, 1.03, and 1.015.

After the manner of the above computation, form a table for the amount of $1 for 50 years at 2%, 2½%, etc., to 10%, and as a number in being multiplied by itself is said to be raised to higher powers, as above,

$$R \times R = R^2,$$
$$R \times R \times R = R^3,$$
$$R^2 \times R^3 = R^5,$$
$$R^4 \times R^3 = R^7.$$

When a number is raised to higher powers, the product has the sum of the powers; therefore, by this table, the compound amount of $1 for any number of years can easily be found, and from the amount subtract 1 and the remainder will be the compound interest.

REM.—If there be a few months or days time above the specified time which cannot be had from the tables, take the highest amount in the tables and compute the balance of time on it as in simple interest, and add it to the amount obtained from the tables.

COMPOUND INTEREST. 221

TABLE

showing the amount of $1, at 2½, 3, 3½, 4, 5, and 6%, compound int., from 1 to 20 years.

Yrs.	2½%.	3%.	3½%.	4%.	5%.	6%.
1	1.025000	1.030000	1.035000	1.040000	1.050000	1.060000
2	1.050625	1.060900	1.071225	1.081600	1.102500	1.123600
3	1.076891	1.092727	1.108718	1.124864	1.157625	1.191016
4	1.103813	1.125509	1.147523	1.169859	1.215506	1.262477
5	1.131408	1.159274	1.187686	1.216653	1.276282	1.338226
6	1.159693	1.194052	1.229255	1.265319	1.340096	1.418519
7	1.188686	1.229874	1.272279	1.315932	1.407100	1.503630
8	1.218403	1.266770	1.316809	1.368569	1.477455	1.593848
9	1.248863	1.304773	1.362897	1.423312	1.551328	1.689479
10	1.280085	1.343916	1.410599	1.480244	1.628885	1.790848
11	1.312087	1.384234	1.459970	1.539454	1.710339	1.898299
12	1.344889	1.425761	1.511069	1.601032	1.795856	2.012197
13	1.378511	1.468534	1.563956	1.665074	1.885649	2.132928
14	1.412074	1.512590	1.618695	1.731676	1.979932	2.260904
15	1.448298	1.557967	1.675349	1.800944	2.078928	2.306558
16	1.484506	1.604706	1.733986	1.872981	2.182875	2.540352
17	1.521618	1.652848	1.794676	1.947901	2.292018	2.692773
18	1.559659	1.702433	1.857489	2.025817	2.406619	2.854339
19	1.598650	1.753506	1.922501	2.106849	2.526950	3.025600
20	1.638616	1.806111	1.989789	2.191123	2.653298	3.207136

EXAMPLES.

1. What is the compound amount of $500 for 12 yrs., at 6%?

The number in the table opposite 12 years, at 6%, is

$2.0121965 \times 500 = \$1006.09825.$

If for 12 years and 6 months,

$2.0121965 \times 1.03 = 2.0725624 \times 500 = \$1036.2812.$

If for 12 years and 3 months, $\times 1.015.$

For any other time, as for a few days, compute the interest on the tabular number as you would in simple interest, and add it to the number of the table.

2. Find the compound amount of $100 for 100 years, at 6%.

The tabular number for 50 years is

$$18.420154 = R^{50} \times R^{50} = R^{100}.$$
$$\underline{18.420154}$$
$$339.3021 \times 100 = \$33930.21.$$

3. Find the compound amount of $100 for 5 yr. 1 mo. 20 da., at 6%.

The tabular amount for 5 years is 1.3382256.

$$1 \text{ mo. } 20 \text{ da.} = 50 \text{ da.}$$
$$\tfrac{50}{6000} = \tfrac{5}{600} = \tfrac{1}{120}.$$
$$1.3382256 \times \tfrac{1}{120} = .0111519.$$
$$1.3382256$$
$$\underline{.0111519}$$
$$1.3493775 \times 100 = \$134.94,$$

amt. for 5 yr. 1 mo. 20 da.

In the above, the interest was payable annually.

4. Find the amount of $500 for 5 years, at 6%, interest payable semi-annually.

The tabular number opposite 3% for 10 years is

$$1.34391638$$
$$\underline{500}$$
$$\$671.95819$$

COMPOUND INTEREST.

The same at 8%, payable quarterly, the tabular number opposite 2% for 20 years is

$$1.4859474$$
$$500$$
$$\overline{\$742.9737}$$

(1.) $\qquad a = P \times R^n.$

The amount is equal to the principal × ratio to the power of the time.

(2.) $\qquad \dfrac{a}{R^n} = P.$

The principal is equal to the amount ÷ ratio to the power of the time.

(3.) $\qquad \dfrac{a}{P} = R^n.$

The ratio to the power of the time = the amt. divided by principal.

FORMULA (1).—Multiply the tabular number opposite time and rate by principal for the amount.

FORMULA (2).—The principal is equal to the amount divided by the tabular number.

FORMULA (3).—To find the rate, divide the amount by the principal, and find the quotient arising therefrom in the table opposite the time, and the rate in which it is found will be the true rate.

To find the time, do the same, and look in the column of the given rate until you find the number opposite the proper time.

5. What principal will produce $2078.928 in 15 years at 5%?

6. At what rate will $450 produce $805.8816 in 10 years? *Ans.* 6%.

7. In what time will $500 produce $1091.44 at 5%? *Ans.* 16 years.

8. In what time will $500 produce $1000 at 6%? That is, in what time will any sum double itself at 6%?

In the table the amount of $1 at 6% for 11 years is 1.8982986, wanting .1017014 to be double. This is $\frac{1017}{18983}$ of the amount = .053574, the rate that 1.8982986 is multiplied by to produce .1017014. The rate at 6% is half the number of months divided by 100.

Hence .053574 × 200 = 10.7148 months.
$$\underline{30}$$
21.4440 days.

Ans. 11 years 10 months and 21 days.

ANNUITIES.

An *Annuity* is a certain sum of money received annually. It may be for a certain fixed time, when it is called a *Certain Annuity*.

It may begin or end with the birth or death of one or more persons, when it is called a *Contingent Annuity*.

A *Perpetual Annuity* is called a *Perpetuity*.

An *Immediate Annuity* begins at once.

A *Deferred Annuity*, or an Annuity in Reversion, begins at a future time.

A *Forborne Annuity* is one in arrears.

ANNUITIES. 225

A sum of money at a given rate of interest produces the same interest every year; this interest is the annuity; hence a perpetuity is the interest of a fixed principal at a fixed rate of interest, and this principal is the present value of the perpetuity.

Let P represent principal or present value.
Let p represent perpetuity.

$$\text{Then (1.) } P \times r = p$$

The perpetuity is the interest of its present worth.

$$(2.)\ P = \frac{p}{r}.$$

The present worth is equal to the perpetuity divided by the rate per cent., and the rate is equal to the perpetuity divided by the present worth.

$$(3.)\ r = \frac{p}{P}.$$

EXAMPLES

1. What perpetuity will a fixed sum of $6000 yield at the fixed rate of 5%?

$$6000 \times \tfrac{5}{100} = \$300.$$

2. What perpetuity will property yield whose fixed value is $12000 at a fixed rate of 6%.

$$\$12000 \times \tfrac{6}{100} = \$720$$

REM.—The present value corresponds to the principal in simple interest, and the perpetuity to the interest of the present value.

3. What is the present value of a perpetuity of $300 a year at 5%.

$$\text{2D Formula. } P = \frac{p}{r}.$$

$$P = \frac{300}{\tfrac{5}{100}} = 300 \times \tfrac{100}{5} = \$6000.$$

4. What is the present value of a perpetuity of $250 at 4%?

$$P = \frac{p}{r} = \frac{250}{\frac{4}{100}} = 250 \times \tfrac{100}{4} = 6250.$$

5. What is the present value of a deferred annuity of $300 at 6%, to commence twenty years hence?

$$300 \times \tfrac{100}{6} = \$5000.$$

As $5000 at 6% produces $300 a year, the present value is such a sum of money as will amount to $5000 in 20 years at 6% compound interest. This sum is the quotient of $5000 divided by the amount of $1 in 20 years at 6%.

Tabular numbers for 20 years at 6% is

3.207136) 5000.0000000 ($1559.02 = present value.

6. What would be the present value of an annuity of $300 at 6%, to begin immediately and continue 20 years?

$5000 − $1559.02 = $3440.98.

7. What is the present value of an annuity of $300 at 6%, to begin in 10 years and then continue 10 years?

Present value of an annuity to begin in 10 years.

1.790848) 5000.00000000 (2791.97

Present value of an annuity to begin
   in 20 years 1559.02

Present value of an annuity to begin
   in 10 years and continue 10 years. $1232.95

ANNUITIES.

8. What is the forborne value of an annuity of $300 for 10 years at 6%?

$5000 is the principal which has been at compound interest 10 years.

The compound interest of $1 for 10 years is

	.790848
And for $5000 is	5000
Is equal to	$3954.240

9. The present value is $6000, interest 6%, payable semi-annually. What annuity will it yield?

Interest on $180 for 6 months,	$5.40
The interest for 1 year,	360.00
	$365.40 annuity.

10. What is the present value of an annuity of $400, payable quarterly; interest at 5%, payable yearly?

$100 × 9 mo. = $100 × 18 mo. at 5% = $7.50
100 × 6 mo.   400
100 × 3 mo.   ─────
                                      407.50 × 20

= $8150.00 = present value.

11. What is the present value of an annuity of $800, payable quarterly; interest also payable quarterly; rate 5%?   Present value = 16000.

REM.—When the payments of the annuity and of the interest on the present value are the same, the computation is the same as if both were annual.

ANNUITIES.

12. Find the present value of a perpetual annuity of $540, beginning ten years hence, rate of interest 6%.
$$\$5025.55 = \text{present value.}$$

13. Find the present value of a certain annuity of $650, beginning at this time and continuing 15 years, rate 5%. $\quad \$6746.78 = $ P. V., C. I.

14. Find the present value of an annuity of $150, to commence in 10 years and then continue 10 years, rate of interest 6%.

Present value of annuity deferred
 10 years 1395.99
Present value of annuity deferred
 20 years 779.51
           ―――
Present value of annuity for 10 yr. $616.48, Pres. value.

15. What is the value of an annuity of $300, unpaid for 12 years, rate 6%?   Ans. $5060.98.

16. A man pays $100 a year for wine and tobacco; what will that amount to at the end of 50 years, allowing 6% compound interest?   Ans $29033.59.

 Rem.—Strict attention must be given to the formulas of annuities, and the students may repeat the examples given until they become familiar with the various formulas.

MENSURATION.

DEFINITIONS.

1. Determining the areas of surfaces and the volume of solids is called *Mensuration.*

2. A *Surface* has two dimensions, as length and breadth; a *Solid* has three dimensions, viz., length, breadth and thickness. A level surface is called a plane.

3. A *Line* has length only.

4. Any position on a surface or a line is a *Point.*

5. The divergence of two straight lines from a point is an *Angle.*

6. If the two straight lines are perpendicular to each other, the angle is called a *Right Angle.*

7. An angle less than a right angle is an *Acute Angle.*

8. An angle greater than a right angle is an *Obtuse Angle.*

9. A plane figure having three angles is a *Triangle.*

10. A plane figure having four angles is a *Quadrilateral.*

11. If the angles of the quadrilateral are right angles it is a *Rectangle,* and if the sides of the rectangle are equal it is a *Square.*

12. If the opposite sides of the quadrilateral are parallel without the angles being right angles, it is a *Parallelogram.*

13. A plane figure of any number of sides has the general name of *Polygon.*

MEASUREMENT OF SURFACES.

PROBLEMS.

1. To find the area of a rectangle, multiply the base by the altitude.

2. To find the area of a parallelogram, multiply the base by the altitude (Geometry).

3. To find the area of a triangle. It is also proved by Geometry that a triangle is one-half a rectangle or parallelogram of the same base and altitude; therefore one-half the product of the base and altitude is the area of a triangle.

EXAMPLES.

1. How many square feet in a board 12 ft. long and 1 ft. wide ? *Ans.* 12 sq. ft.

2. How many square feet in a board 12 ft. long and 10 in. wide ?
$$12 \times \tfrac{10}{12} = 10 \text{ sq. ft.}$$

3. How many square feet in a board 12 ft. long and 9 in. wide ?
$$12 \times \tfrac{9}{12} = 9 \text{ sq. ft.}$$

MENSURATION. 231

4. How many square feet in a room 18 ft. long and 12 ft. wide?

$$18 \times 12 = 216 \text{ sq. ft.}$$

5. How many square yards in a room 18 ft. long and 12 ft. wide?

$$\frac{\overset{2}{18 \times 12}}{\underset{1}{9}} = 24 \text{ sq. yd.}$$

6. How many square feet in a room 16 ft. 9 in. long and 12 ft. 9 in. wide?

$$\frac{\overset{67}{201}}{\underset{1}{12}} \times \frac{\overset{25}{151}}{\underset{4}{12}};$$

$$\therefore \frac{67 \times 25}{2 \times 4} = \frac{1675}{8} = 209\tfrac{3}{8} \text{ sq. ft.}$$

7. How many square feet in a wall 18 ft. 9 in. long and 9 ft. 6 in. high?

$$\frac{\overset{25}{225}}{\underset{4}{12}} \times \frac{\overset{57}{114}}{\underset{2}{12}} = \frac{25 \times 57}{8} = \frac{1425}{8} = 178\tfrac{1}{8} \text{ sq. ft., } Ans.$$

8. How many square feet in a wall 16 ft. long and 9 ft. 6 in. high?   *Ans.* 152 sq. ft.

9. How many square feet in the four walls of a room having the dimensions of 7th and 8th examples?

Ans. 660¼ sq. ft.

10. What number of square feet in the floor or ceiling?

$$\frac{\overset{75}{225}}{\underset{2}{12}} \times \overset{4}{16} = 300 \text{ sq. ft. in each.}$$

11. How many square feet of plastering, in the walls and ceiling of this room? How many square yards?

1st Ans. 960¼ sq. ft.   *2d Ans.* 106²⁴⁄₃₆ sq. yd.

12. How many square feet in the roof of a house 50 ft. long and the rafters on each side of the roof 25 ft. long?

$$50 \times 50 = 2500 \text{ sq. ft.}$$

13. How many shingles, each shingle covering 6 in. by 4, will it take to roof the house?

$$\frac{2500 \text{ sq. ft.} \times \overset{6}{\cancel{144}}}{\cancel{24}} = 15000 \text{ shingles.}$$

14. How many square yards of plastering in a house 20 ft. front, 60 ft. deep and 36 ft. high; three stories in height? and how many shingles to roof it, each covering 8 in. by 6, and the entire length of the roof 64 ft., the width 20 ft.?   *1st Ans.* 1040 sq. yd. plastering.
    2d Ans. 3840 shingles.

15. What is the area of a triangle whose base is 24 in. and the perpendicular distance from the vertical angle to the base 9 in.?

$$\frac{\overset{12}{\cancel{24}} \times 9}{\cancel{2}} = 108 \text{ inches.}$$

MEASUREMENT OF CIRCLES.

DEF.—A *Circle* is a plane figure bounded by a curved line called the circumference, every point of which is equally distant from a point within, called the centre.

PROBLEMS.

1. To find the circumference of a circle, having given the diameter. Multiply the diameter by 3.1416 (Geometry).

MENSURATION. 233

2. To find the diameter, having given the circumference. Divide the circumference by 3.1416.

3. To find the area of a circle. Multiply the circumference by one-fourth the diameter.

Let D = diameter,
then $D \times 3.1416$ = circumference,
and $D \times \overset{.7854}{3.1416} \times \dfrac{D}{4} = D^2 \times .7854$;

or, multiply the square of the diameter by .7854.

EXAMPLES.

1. Find the circumference of a circle whose diameter is 2 ft. *Ans.* 6.2832 ft.

2. Find the diameter of a circle whose circumference is 6.2832 ft. *Ans.* 2 ft.

3. Find the area of a circle whose diameter is 2 ft.

$$2 \times 2 \times .7854 = 3.1416 \text{ sq. ft.}$$

PROB. 4.—Find the space between two concentric circles. Find the area of both circles, and the difference of the areas will be the area of the space.

Ex. 4. Two circles have the same centre; the larger one has a diameter of 4 feet and the smaller one of 2 ft. What is the area of the space between them?

$$\begin{array}{r} 4 \times 4 = 16 \\ 2 \times 2 = \underline{4} \\ 12 \times .7854 = 9.4248 \text{ sq. ft.} \end{array}$$

CUBIC MEASURE.

DEFINITIONS.

1. A *Right Prism* is a solid which has equal and parallel polygons for its bases, and its edges are perpendicular to the bases.

2. If the bases are squares and each face square and equal to a base, the figure is called a *Cube.*

3. When the bases are circles, the figure becomes a *Cylinder.*

4. Figures with a polygon for a base and tapering to a point are called *Pyramids.*

5. If the upper part is cut off by a plane parallel to the base, the lower part is called the *Frustum* of a *Pyramid.*

6. When the base is a circle and the figure tapers to a point, it is called a *Cone.*

7. If the upper part is cut off by a plane parallel to the base, the lower part is called the *Frustum* of a *Cone.*

PROBLEMS.

To find the volume of a prism or cylinder.

Multiply the area of the base by the altitude. (Geom.)

If it be a cube, the cube of an edge will be the volume.

REM.—The lateral surface of a prism or cylinder is the product of the altitude and the perimeter of the base.

MENSURATION. 235

EXAMPLES.

1. What is the volume of a prism whose base contains 6 sq. ft. and altitude 5 feet?

$$6 \text{ sq. ft.} \times 5 \text{ ft.} = 30 \text{ cu. ft.}$$

2. What is the volume of a cube whose edges are each 3 feet? $3 \times 3 \times 3 = 27$ cu. ft.

3. What is the volume of a cylinder whose base contains 12 sq. ft. and its altitude is 5 ft.

$$12 \text{ sq. ft.} \times 5 \text{ ft.} = 60 \text{ cu. ft.}$$

PROBLEM.

To find the volume of a pyramid or cone.

Multiply the area of the base by one-third the altitude.

EXAMPLES.

1. What is the volume of a pyramid whose base contains 15 sq. ft. and altitude 9 ft. ?

$$15 \times 3 = 45 \text{ cu. ft.}$$

2. The volume of a pyramid is 90 cu. ft. and the altitude 9 ft.; what is the area of base? *Ans.* 30 sq. ft.

3. What is the volume of a cone whose base contains 36 sq. ft. and its altitude is 18 ft.? *Ans.* 216 cu. ft.

PROBLEM.

To find the volume of the frustum of a pyramid or cone.

The volume of the frustum is equivalent to three pyramids or cones, one having the lower base for its base, the second having the upper base for its base, each having for its altitude the altitude of the frustum, and the volume of the third pyramid or cone is a mean proportional between the other two.

Therefore, extract the square root of the product of the areas of the lower and upper bases, and add together this root and the areas of the two bases and multiply their sum by one-third the altitude of the frustum, and this product will be the volume of the frustum.

EXAMPLES.

1. The areas of the lower and upper bases of the frustum of a pyramid are 16 sq. ft. and 9 sq. ft., and the altitude is 12 ft.; what is the volume of the frustum.

$$16 \times 9 = \sqrt{144} = 12$$
$$16$$
$$\underline{9}$$
$$37 \times 4 = 148 \text{ cu. ft.}$$

DEFINITIONS.

1. A *Sphere* is a solid with a curved surface, every part of which is equally distant from a point within called the centre.

2. The *Axis* or *Diameter* of a sphere is a straight line passing through the centre and terminated at both ends by the surface.

3. A *Radius* is one-half the diameter, or a straight line drawn from the centre to any point of the surface.

MENSURATION.

PROBLEMS.

I. To find the surface of a sphere.

The surface of a sphere is equal to the product of the circumference and diameter.

$\pi \times D \times D$ = surface.   (Geometry.)

Cor.—The surfaces of spheres are to each other as the squares of their diameters.

EXAMPLES.

What is the surface of a sphere whose diameter is 3 ft.?

Circumference = 3.1416×3.

Surface = $3.1416 \times 3 \times 3 = 28.2744$ sq. ft.

II. To find the volume of a sphere.

The product of the surface and $\frac{1}{3}$ the radius or $\frac{1}{6}$ the diameter.

$\pi = 3.1414$ and D = diameter.

Volume = $\pi \times D \times D \times \dfrac{D}{6} = .5236 \times D^3$.

The volume of a sphere is equal to the cube of the diameter multiplied by .5236.

Cor.—The volumes of spheres are to each other as the cubes of their diameters.

EXAMPLES.

1. What is the volume of a sphere whose diameter is 2 feet? $.5236 \times 8 = 4.1888$ cu. ft.

2. How many shot $\frac{1}{16}$ of an inch in diameter may be made of a ball of lead 6 inches in diameter?

$\dfrac{6 \times 6 \times 6}{\frac{1}{16} \times \frac{1}{16} \times \frac{1}{16}} = 6 \times 6 \times 6 \times 16 \times 16 \times 16 = 884736$ shot.

ANSWERS NOT GIVEN

IN THE PRECEDING PAGES.

Addition and Subtraction.

Page 18.

1. 3 remain in first field, and there are then 12 in the second.
2. 115 remain in first field, and there are then 977 in the second.
3. $2365.   4. $323.   5. $27586.

Page 19.

6. $57.
7. $5220.
8. Lost $194.
9. $379.
10. $175321.
11. $36.
12. 36311 yds.
13. $131179.

Multiplication.

Page 25.

1. 18796796.
2. 12895164.
3. 32671178.
4. 17600100.
5. 505489116.
6. 3515971700.

7. 1913247450. 9. 982259375.
8. 8585904128. 10. 984381300.

DIVISION.

Page 30.

1. Quo. 1915, rem. 98.   7. Quo. 800368, rem. 209.
2. "   1477, "   60.   8. "   373792, "   979.
3. "   263, " 234.   9. "   439169, " 1458.
4. "   5475, " 116.   10. "   13482, "   42.
5. " 209205, " 992.   11. "   4457, " 102170.
6. " 780317, " 300.   12. "   676, " 196432.

   1. 50 cts.   4. 10 lbs.   7. 200 cts.
   2. 5 lbs.   5. 150 cts.   8. 20 lbs.
   3. 100 cts.   6. 15 lbs.   9. 250 cts.
                  10. 25 lbs.

Page 31.

11. $72.   16. $492.   21. $2085.
12. $135.   17. $3488.   22. $8.
13. 225 cts.   18. $2592.   23. $160.
14. 693 cts.   19. 387 trees.   24. 120 acres.
15. 2352 cts.   20. 252 yards; cost, $1512.

Page 32.

25. 20 acres.   29. 10 heads.   32. $52239.
26. 15 years.   30. 8 acres.   33. $966.
27. $2.   31. $247640 to grandson.
28. 25 rows.
34. Carriage, $513; $63 more than the horses.

Page 33.

35. I gained $400.
36. 67 years.
37. 1570 years before Christ.
38. 60 cts.
39. 165 cts.
40. 3015 cts.
41. $322.
42. $18705.
43. 26355 cts.
44. 12 cts.
45. 11 cts.
46. 335 lbs.

Page 34.

47. 23 tons.
48. 435 acres.
49. 35 cts.
50. 144 cts.
51. 880 cts.
52. 324 cts.
53. $966.
54. $3741.
55. 8785 cts.
56. $56250.
57. $7856.

FACTORING.

Page 38.

17. $51 = 3, 17.$
18. $52 = 2, 2, 13.$
19. $54 = 2, 3, 3, 3.$
20. $55 = 5, 11.$
21. $56 = 2, 2, 2, 7.$
22. $57 = 3, 19.$
23. $58 = 2, 29.$
24. $60 = 2, 2, 3, 5.$
25. $62 = 2, 31.$
26. $63 = 3, 3, 7.$
27. $64 = 2, 2, 2, 2, 2, 2.$
28. $65 = 5, 13.$
29. $68 = 2, 2, 17.$
30. $70 = 2, 5, 7.$
31. $72 = 2, 2, 2, 3, 3.$
32. $74 = 2, 37.$
33. $75 = 3, 5, 5.$
34. $76 = 2, 2, 19.$
35. $77 = 7, 11.$
36. $78 = 2, 3, 13.$
37. $80 = 2, 2, 2, 2, 5.$
38. $81 = 3, 3, 3, 3.$
39. $82 = 2, 41.$
40. $84 = 2, 2, 3, 7.$
41. $85 = 5, 17.$
42. $86 = 2, 43.$

DECIMAL FRACTIONS.   241

43. 87 = 3, 29.
44. 88 = 2, 2, 2, 11.
45. 90 = 2, 3, 3, 5.
46. 91 = 7, 13.
47. 92 = 2, 2, 23.

48. 94 = 2, 47.
49. 105 = 3, 5, 7.
50. 210 = 2, 3, 5, 7.
51. 336 = 2, 2, 2, 2, 3, 7.
52. 540 = 2, 2, 3, 3, 3, 5.

Page 43.

EXAMPLE 4. 50 yards.

Page 54.

9. $2062\frac{1}{4}$.
10. $\frac{12317}{35} = 3445\frac{1}{35}$.
11. $5202\frac{1}{13}$.
12. $4418715\frac{3}{4}$.
13. $1491263\frac{4}{15}$.

Page 57.

3. 152.   4. 202.   5. 1.

Page 60.

5. $\frac{3}{4} \times \frac{5}{11} = \frac{3}{4} \times \frac{5}{7} \times \frac{7}{5} \times \frac{\overset{3}{\cancel{12}}}{11} = \frac{9}{11}$.

6. $\frac{2}{3} \times \frac{4}{5} = \frac{2}{3} \times \frac{4}{5} \times \frac{7}{6} \times \frac{8}{7} = \frac{32}{45}$.

DECIMAL FRACTIONS.

Page 66.

Observe that in the first example, 1 is a factor of each product; hence the product is the same as the other factor.

In the second example, one of the factors is one-tenth, consequently the product is one-tenth of the other factor; that is, consists of the same figures, with one additional decimal figure, and this another zero placed before the product of the first examples.

1. .1, .01, .001, .0001.
2. .01, .001, .0001, .00001.
3. .4, .012, .0020, .00018, .000012.
4. 1.63254.
5. .043554.   8. 15455.8152.
6. .007082.   9. 1389171.48.
7. .0007508.   10. 3506284.8.

Page 68.

8. Quotient, 4525.06; remainder, 16810.
9. Quotient, 84.8; remainder, 62121.
10. Quotient, 19; remainder, 5246885.

REM.—The division may be carried as far as necessary by annexing zeros to the dividend, as every additional decimal in the dividend will give an additional one in the quotient.

DENOMINATE NUMBERS.

Page 77.

4. £291 15s. 9d.

Page 78.

2. 500 cts. 5000 mills.   5. 6153 mills.
3. 700 cts. 7000 mills.

DENOMINATE NUMBERS. 243

Page 79.

9. 63 dollars 25 cents 7 mills.
10. 753 dollars 25 cents.
11. $9.   13. $105.   15. $39.375.
12. $10.50.   14. $1.125.   16. $55.65.
17. Cost $483.136; sold for $640.328.
18. 60 cts., 70, 80, 90, 100, 110, 120.
19. 84 cts., 96, 108, 120, 132, 144.
21. 4½d., 5¼, 5¾, 6¼, 6¾, 7¼, 7¾, 8¼, 8½, 8¾.
22. 2$\frac{1}{13}$s., 2$\frac{4}{13}$, 2$\frac{11}{13}$, 3$\frac{8}{13}$, 3$\frac{9}{13}$, 4$\frac{3}{13}$, 4$\frac{9}{13}$, 5$\frac{5}{13}$.
23. £1$\frac{11}{20}$, 2, 2$\frac{19}{20}$, 3, 3$\frac{5}{20}$, 3$\frac{19}{20}$, 3$\frac{15}{20}$, 4, 4$\frac{5}{20}$, 4$\frac{19}{20}$, 4$\frac{15}{20}$, 5, 5$\frac{5}{20}$, 5$\frac{19}{20}$, 6.

Page 80.

24. 9303 farthings.   26. £39 19s. 3d. 3 far.
25. £39442 0s. 0d. 1 far.   27. £1 19s. 2d. 0$\frac{9}{13}$ far.

Page 89.

1. 3359 far.   12. 1 lb. 2 ℥ 0 ʒ 1 ℈ 2 gr.
2. 3068 pence.   13. 741402 inches.
3. 1682 pence.   14. 4 lea. 0 mi. 5 fur. 12
4. £2 9s. 0d. 2 far.   rd. 4 yd. 1 ft. 4 in.
5. £2 5s. 3d.   15. 177¼ in.
6. £36 11s.   16. 135 inches.
7. 1705577 drams.   17. 180 inches.
8. 1 T. 6 cwt. 1 qr. 1 lb. 11 oz.   18. 162 inches.
9. 32871 gr.   19. 182416 sq. in.
10. 11 lb. 2 oz. 0 pwt. 4 gr.   20. 161940 cu. in.
11. 14690 gr.   21. 2 cords.

Page 90.

22. 831 gills. 23. 633 pt.

Page 93.

56. 1440 yd.; 1200 yd.; 960 yd.; 640 yd.; 480 yd.; 384 yd.; 320 yd.; 274$\frac{2}{7}$ yd.

57. Proceeds of flour, $3312.50.
Number of yards, 3785$\frac{5}{7}$.

RATIO.

Page 103.

9. $\cancel{\$5.40}^{.60} \times \dfrac{2}{\cancel{9}} = \$1.20$ per bushel, then $\$1.20 \times \cancel{\dfrac{35}{4}} = \$10.50$; $\cancel{\$1.20}^{.60} \times \dfrac{19}{\cancel{2}} = \$11.40$; $28.50; $38.00; $57; $47; $70.50; $117; $128.10.

10. 11.87\frac{1}{2}$; 12.81\frac{1}{4}$; 19\frac{1}{2}$; 46\frac{1}{2}$; 119\frac{1}{2}$; $157; 188\frac{1}{2}$; 349\frac{1}{4}$.

11. $260; $310; $500; $525; $690; $735; $780; $825; $1510; $1995.

12. 37\frac{1}{2}$; $25; 12\frac{1}{2}$; $10; $20; $30; $40; $60; $70.   13. 1\frac{11}{15}$.

Page 104.

14. $9.   15. $18.75, $22.50.

Page 105.

7. 15 barrels.   11. $3500.   13. $49.50.
8. 15 days.   12. $42.60.

PERCENTAGE. 245

Page 106.

15. $1242.15.   21. 146 lb.   25. $2075.
17. $822.79+.   22. $94.25.

Page 107.

26. $5445.   27. $112.   30. $36.36.

Page 112.
EXAMPLE 9. 11 men.

PERCENTAGE.

Page 123.—Ex. 8. $3921.57.

Page 124.

13. $1860.48.   15. $108000.   17. $3851.69+.
14. $7760.   16. $4901.96+.

Page 127.—Ex. 10. $900.94.

Page 136.—Ex. 8. $96.37.

Page 141.

3. $983.61.   9. 7%.
6. $4800.   12. 2 yr. 2 mo. 12 da.

Page 147.—Ex. 6. $1000.   Ex. 7. $1000.
Page 149.—Ex. 5. $25979\frac{31}{34}$ francs.   Ex. 6. $5000.
Page 150.—Ex. 10. $5000.

EVOLUTION.

Page 162.

1. 99.   2. 321.   3. 444.   4. 999.   5. 777.

Page 163.—Ex. 2.   17, nearly.   Ex. 3.   75.

SERIES OF EQUAL RATIOS.

Page 174.—Ex. 4.   3280.

COMPOUND INTEREST.

Page 177.

To find a number greater than any given in the table.

As $R^2 \times R^2 = R^4$, hence, the product of any two tabular numbers will be the tabular number for the sum of the years corresponding to those numbers.

Thus, tab. No. opp. 20 years × tab. No. 20 × tab. No. 10 = tab. No. opp. 50 years.

Tab. No. of 20 × tab. No. 9 = tab. No. 29, etc.

Page 180.—Ex. 5. $1000.

THE METRIC SYSTEM.

The Metric System is a decimal system of measures and weights used in France and many other countries. It corresponds to our decimal system and is likely to become general, as the benefits of a general system are obvious.

The units of this system are the Metre, the Are (pronounced air), the Litre (pronounced leetur), the Gram, and the Stere (pronounced Stare).

The unit holds a central position as in abstract numbers, and the orders to the left increase by tens, and are designated by the prefixes deca, hecto, kilo, and myria, whilst the decimals to the right diminish by tenths, hundredths, etc., and are designated by the prefixes deci, centi, and milli.

The unit of length is the Metre from which the name of the system is derived, and its length is $\frac{1}{10000000}$ of an arc of 90 degrees or $\frac{1}{4}$ the circumference of the earth, and equals 39.37079 inches.

MEASURES OF LENGTH.

10 millimetres	equal	1 centimetre	=	.3937079
10 centimetres	"	1 decimetre	=	3.937079
10 decimetres	"	1 metre	=	39.37079 = 3 ft. 3.37079 in.
10 metres	"	1 decametre	=	393.7079
10 decametres	"	1 hectometre	=	3937.079
10 hectometres	"	1 kil'ometre	=	39370.79
10 kilometres	"	1 myr'iametre	=	393707.9

The kilometre is the unit for expressing long distances.

The unit of Surface is the are, which is a square decametre, or a square whose side is 10 metres.

MEASURES OF SURFACE.

The cen'tiare	=	1 square metre.
100 centiares	=	1 are.
100 ares	=	1 hectare.

The hectare is the unit of measure for land.

The unit of capacity is the Litre, which is equivalent to a cube whose edge is $\frac{1}{10}$ of a metre, and is a little more than a quart liquid measure.

MEASURES OF CAPACITY.

10 mil'lilitres	=	1 cen'tilitre.
10 centilitres	=	1 decilitre.
10 decilitres	=	1 Litre.
10 Litres	=	1 dec'alitre.
10 decalitres	=	1 hectolitre.
10 hectolitres	=	1 kil'olitre.

The unit of Liquid Measure is the Litre, and the hectolitre of Dry Measure.

In the measurement of wood, the kilolitre or cubic metre is the unit, and is called the Stere; it equals .2759 of a cord, and 10 steres = 1 dec'astere = 2.759 cords.

WEIGHTS.

10 milligrams	=	1 centigram.
10 centigrams	=	1 decigram.
10 decigrams	=	1 Gram = nearly 15¼ gr.
10 Grams	=	1 decagram.
10 decagrams	=	1 hectogram.
10 hectograms	=	1 kilogram.
10 kilograms	=	1 myriagram.
10 myriagrams	=	1 quintal.
10 quintals	=	1 tonneau.

The kilogram is the unit in general dealings, but in very large quantities the tonneau is used.

We numerate the several denominations as we do abstract numbers, regarding but the one name, as Metre, Are, Litre, and Gram, or by the unit applied to the particular case, and all computations are the same as abstract numbers or U. S. money.

TEST EXAMPLES.

1. The sum of two numbers is 504, and one of the numbers is 253; what is the other number? *Ans.* 251.

2. The sum of two numbers is 753, and the one is 51 more than the other; what are the numbers?

Ans. 351 and 402.

3. The product of two numbers is 255, and one of the numbers is 17; what is the other number?

Ans. 15.

4. The product of three factors is 3276; two of the factors are 12 and 21; what is the other factor?

Ans. 13.

TEST EXAMPLES.

5. The divisor is 14, the quotient 35, and the remainder 5; what is the dividend? *Ans.* 495.

6. What is the quotient of 65 bu. 1 pk. 3 qt. divided by 12? *Ans.* 5 bu. 1 pk. $6\frac{1}{4}$ qt.

7. In 7960 farthings, how many pounds?
Ans. £$8\frac{7}{24}$.

8. In £9 10s. 8d. 3 far., how many farthings?
Ans. 9155 far.

— 9. Multiply 3 years 7 months and 20 days by 5.
Ans. 18 y. 2 m. 10 d.

10. To what term in division does the numerator of a fraction correspond? the denominater? the fraction itself?

11. In the multiplication of decimals, how does the number of decimals in the product compare with that of the factors?

12. What is the quotient of $\frac{9}{17}$ divided by 3?
Ans. $\frac{3}{17}$.

13. What is the product of a fraction multiplied by its denominator? Give an example.

14. In the division of decimals, how does the number of decimals in the dividend compare to that of the divisor and quotient?

15. What do you do when the divisor has more decimals than the dividend?

16. When there is a remainder in the division of decimals, can you form a common fraction with it and the divisor, as in the division of integers?

17. When the dividend and divisor have the same number of decimals, how may they be regarded?

18. How do you reduce an integral number to a fraction of any given denominator?

19. How do you reduce a fraction to its lowest terms?

20. How do you reduce fractions to the least common denominator?

21. Reduce ½, ⅓, ¼, ⅕ and ⅙ to the least common denominator, and then find their sum.
 Ans. Com. Den., 60; sum, 3 11/20.

22. What is the difference between ½ and ⅓?
 Ans. 1/6.

23. Multiply ½ by ¼ and explain the process. *Ans.* ⅛.

24. Divide the same and explain the process.
 Ans. ⅔.

25. Is it necessary that fractions have a common denominator in order to add or subtract them? Is it necessary in order to multiply them? To divide them?

26. How many seconds in a year of 365 days?
 Ans. 31536000 sec.

27. How many seconds in the circumference of a circle? *Ans.* 1296000″.

28. Is there any difference in the number of seconds in a large or a small circle?

29. The length of the front building of a house is 54 feet and the width 36 ft., the length of the wing is 48 ft. and the width 18 ft.; what is the greatest length of boards with which to weatherboard it without cutting any of them? *Ans.* 6 feet.

30. Put .375 in the form of a common fraction, and reduce it to its lowest terms. *Ans.* ⅜.

31. Is any even number a prime number?

32. What is the shortest method of dividing by a fraction?

33. What is the shortest method of multiplying a number of fractions?

34. How does the product of any number of proper fractions compare with any one of its factors? How of improper fractions?

35. How does the quotient compare with the dividend, when the divisor is a proper fraction? How when the divisor is an improper fraction?

36. If $\frac{7}{8}$ of a yard of cloth cost 63 cents, what will a yard cost? *Ans.* 72 cts.

37. If $\frac{5}{8}$ of an acre of land cost \$75, what will an acre cost? *Ans.* \$120.

38. If $\frac{3}{4}$ of a farm is worth \$4200, what is $\frac{5}{7}$ worth? *Ans.* \$4000.

39. If $\frac{4}{5}$ of an acre is worth \$40, what is $\frac{7}{10}$ worth? *Ans.* \$35.

40. If $\frac{5}{8}$ of a yard cost $\$3\frac{3}{4}$, what will $\frac{3}{4}$ of a yard cost? *Ans.* $\$4\frac{1}{2}$.

41. If $\frac{3}{16}$ of a farm cost \$8769, what will $\frac{5}{8}$ of it cost? *Ans.* \$29230.

42. What is the rate per hour of a boat that goes $312\frac{4}{5}$ miles in $33\frac{1}{3}$ hours? *Ans.* $9\frac{48}{125}$ miles.

43. In a garrison 480 men have provisions for 9 months; how many men must leave at the end of six months that the remaining provisions may last six months longer? *Ans.* 240 men.

44. Divide 369 into three parts that shall be to each other as 2, 3, and 4. *Ans.* 82, 123 and 164.

45. Divide 1728 into four parts in the ratio of 2, 3, 4, and 9. *Ans.* 192, 288, 384, 864.

46. One-third is what per cent?

47. One-ninth is what per cent?

48. Two-ninths is what per cent?

49. Forty is how many per cent of 200?

TEST EXAMPLES.

50. Fifty is how many per cent of 400 ?

51. A note for $5000 dated April 1st, 1872, with interest at 6%, is endorsed as follows: May 1st, 1873, received $1025, June 1st, 1874, $1020, July 1st, 1875, $1015. What was due August 1st, 1876 ?   *Ans.* $2956.30.

52. What is the bank discount of a note for $600 payable in sixty days, interest at 6%, 3 days grace? and what are the proceeds?
 Ans. Discount $6.30, and Proceeds $593.70.

53. What is the bank discount of $600 for 3 months at 6% ?   *Ans.* $9.30.

54. What is the amount of $600 for 3 years and 4 months at 6 per cent simple interest? At compound interest?   *Ans.* $720, and $728.90.

55. A person owning ⅝ of a mine, sells ¾ of his share for $5130; what was the whole mine worth at that rate?
 Ans. $10944.

56. What is the cost of 15 T. 12 cwt. 2 qr. 10 lb. coal, at $6 per ton?   *Ans.* $93.78.

57. How is a common fraction changed into a decimal?

58. How many terms form a ratio?

59. How many ratios may be formed of any two numbers either integral or fractional?

60. What is the present value of the following note to be discounted in bank, interest at 6%?

Ninety days after date, I promise to pay to the order of John Smith, three hundred dollars, value received.
 THOS. BROWN.
 Ans. $295.35.

61. Extract the square root of .00012544.
 Ans. .0112.

62. How many steps of three feet each would a man take in walking a mile?

63. By what process would 25643.21 be changed to 256.4321?

64. Reduce .625 lb. to ounces. *Ans.* 10 oz.

65. Reduce 12 lb. 8 oz. 4 drams to the decimal of cwt. *Ans.* .125+.

66. Solve the following by proportion, and explain the solution: If 27 acres of land cost $1350, what will 36 acres cost at the same rate? *Ans.* $1800.

67. How many cords of wood can be put in a shed 256 ft. long 20 ft. wide and 8 ft. high? *Ans.* 320 cords.

68. How many shingles netting 6 in. by 4, will cover a house 50 ft. long and 40 ft. wide, with gable roof and the length of rafter 25 ft. *Ans.* 15000 shingles.

69. How many square ft. of lumber in one floor of the above house? *Ans.* 2000 sq. ft. And how many square yards of plastering in a hall 15 feet high extending over the whole building? *Ans.* 522⅔ sq. yds. How many yards of carpeting one yard wide will cover the floor? *Ans.* 222⅔ yds.

70. The product of two equal factors is 34225; what is each factor? Solve by factoring. By what other method may it be solved?

71. What is the greatest common divisor of 72 and 1728? *Ans.* 72.

72. Find the least common multiple of 6, 9, 12, 18, and 36. *Ans.* 36.

73. Reduce ⅛ lb. to the decimal of a ton. *Ans.* .0000625 ton.

74. What is the amount of $200 for 3 years at 6% compound interest, payable semi-annually? *Ans.* $238.81.

TEST EXAMPLES.

75. If 12 men build a wall 36 yds. long in 30 days, working 8 hours a day, how many men can build a wall 84 yds. long in 60 days, working 7 hours in a day ?
Ans. 16 men.

76. What is the length of a side of a cubical box containing 48228544 cubic inches ? *Ans.* 364 in.

77. Extract the square root of $\frac{4}{7}$. *Ans.* .745+.

78. In a series of equal ratios, the first term is 3, the ratio 2, and the last term 192 ; what is the sum of the series ? *Ans.* 381.

79. A gentleman giving his daughter in marriage on New Year's day, gave her husband $100 toward her portion, promising to double it on the first day of every succeeding month during the year; what was her portion ? *Ans.* $409500.

80. If a man travel 540 miles in 30 days of 9 hours each, how far will he travel in 48 days of 12 hours each ?
Ans. 1152 miles.

81. How many numbers form a proportion ?

82. What are the names of the terms ?

83. How many terms form a ratio ? and what are they called ?

84. The two means of a proportion are 3 and 6, and one of the extremes is 2 ; what is the other extreme ?
Ans. 9.

85. The two extremes of a proportion are 5 and 9, and one of the means is 3 ; what is the other mean ?

86. In a series of ratios forming proportions, what name is given to the odd terms, and what to the even ones ?

87. In the usual method of numeration, how many figures in each period ?

TEST EXAMPLES.

88. Name the first three periods of integers, and the first four places of decimals.

89. How does multiplying the numerator of a fraction affect the value of the fraction? How multiplying the denominator?

90. How does a zero annexed to a decimal affect the value? If a zero is prefixed what is its effect?

91. A father left to his four sons $38750, to be divided in such a manner that each one's share being placed at 5% simple interest shall amount to the same when he shall become 21 years of age, their ages at his death being 13, 15, 17 and 19 years respectively; what principal must be placed at interest for each?

Ans. $8580 youngest; $9240, $10010, $10920 oldest.

The ratios will be

$\frac{1\cdot 8}{4\cdot 0}$, $\frac{1\cdot 8}{3\cdot 0}$, $\frac{1\cdot 8}{2\cdot 0}$, $\frac{1\cdot 8}{1\cdot 0}$

$\frac{1\cdot 2}{4}$, $\frac{1\cdot 2}{3}$, $\frac{1\cdot 2}{2}$, $\frac{1\cdot 2}{1}$

4290
4620
5005
5460

19375

$\frac{4}{1}$, $\frac{14}{3}$, $\frac{4}{1}$, $\frac{14}{1}$ = $\frac{444}{...}$, $\frac{444}{...}$, $\frac{444}{...}$, $\frac{444}{...}$

$\frac{4290}{19375} \times \$38750 = \$8580$, etc.

Reduce to a common denominator, and the numerators express the ratios.

92. The difference of time between Washington and Cincinnati is 29 min. 36 sec.; what is the difference of longitude? *Ans.* 7° 24′.

93. The difference of time between two places is 2 hours 20 min. 40 sec.; what is the difference of longitude?

Ans. 35° 10′.

94. The difference of time between Baltimore and

TEST EXAMPLES.

New Orleans is 53 min. 30 sec.; what is their difference of longitude? *Ans.* 13° 22′ 30″.

95. A strip of land is 10 rods wide; how long a piece must be cut off from it to make just ¼ of an acre?
Ans. 12 rds.

96. The diameter of a circle is 10 ft.; what must be the diameter of another circle to contain just ten times the area of the first? *Ans.* 31.62 feet.

97. The diameter of a ball weighing 10 lb. is 2 inches; what must be the diameter of a ball of the same material to weigh ten times as much? *Ans.* 4.3 inches.

98. A merchant bought a lot of cloth which he marked ¼ above cost. Owing to a sudden fall in prices he was obliged to sell it at one dollar a yard, which was one-third off the marked price; what did he lose per yard? *Ans.* 25 cts.

99. Marked goods at an advance of 20% on the cost price; if I take off 20% from the marked price, do I gain or lose, and what per cent? *Ans.* I lose 4%.

100. Two men bought a barrel of flour for $9; one paid $4¼ and the other the rest; what part of the flour ought each to get for his share?
Ans. The 1st $\frac{17}{36}$, and the other $\frac{19}{36}$.

101. Galveston is 14° 43′ west of Pittsburg. When it is noon at Galveston, what is the time at Pittsburg?
Ans. 58 min. 52 sec. P. M.

102. The longitude of Rome is 20° 30′ east and the longitude of Baltimore is 76° 37′ west. When it is 9 A. M. at Rome, what time is it at Baltimore?
Ans. 2 h. 31 m. 32 sec. A. M.

103. Sold tea at 90 cts. a pound and gained 20%; what % could be gained by selling at $1 a lb.? *Ans.* 33⅓%.

104. A merchant retails goods at 30% above cost, but sells them by wholesale at 12% below the retail price; what does he gain per cent on goods sold at wholesale?
Ans. 14$\frac{2}{5}$%.

105. A merchant marked a piece of silk at 25% above cost, and then sold it at 20% below the marked price; did he gain or lose? *Ans.* Neither.

106. How many shares of bank stock ($100) at 4% premium can be bought for $8320? *Ans.* 80 shares.

107. How much railroad stock at 12$\frac{1}{2}$% discount, can be bought for $8750? *Ans.* 100 shares.

108. A broker bought stock at 4% discount, and selling the same at 5% premium, gained $450; how many shares did he buy? *Ans.* 50 shares.

109. For what must a cargo valued at $11520 be insured at 4% to cover both goods and premium?
Ans. $12000.

110. A sold cloth to B and gained 10%; B sold it to C and gained 10%; C sold it to D for $726 and gained 10%; how much did it cost A? *Ans.* $545.454.

111. A and B together did $\frac{9}{10}$ of a piece of work in 2 days, when, B leaving, A completed it in $\frac{1}{2}$ a day; in what time can each do it alone?
Ans. A in 5 days, B in 4 days.

112. In a company of 90 persons, there are 4 more men than women, and 10 more children than men and women together; how many of each in the company?
Ans. 18 women, 22 men, 50 children.

113. How many bushels of wheat will fill a bin 8 ft. long, 5 ft. wide, and 4 ft. deep? *Ans.* 128.57+ bu.

114. How many square feet on the surface of a block 6 ft. long, 4 ft. wide, and 1$\frac{1}{2}$ ft. thick? *Ans.* 78 sq. ft.

TEST EXAMPLES.

115. There is a circular field 40 rods in diameter; what is its circumference and how many acres does it contain? *Ans.* Circum. 125.664 rds.; area 7.854 acres.

116. How large a draft may be purchased for $2020, at 1% premium? *Ans.* $2000.

117. For what sum must a note be drawn so that when discounted for 63 days at 6%, the proceeds may be $1295? *Ans.* $1308.74.

118. Invested $6000 at 6% and received $750.42 as interest; how long was my money invested? *Ans.* 2.0845 yrs.

119. A, B and C join their capitals, which are in the proportion of $\frac{1}{2}$, $\frac{1}{3}$ and $\frac{1}{4}$. At the end of 4 months A withdraws one-half of his capital, and at the end of 9 months more they divide their profits, $1420; what would each receive? *Ans.* A, $510; B, $520; C, $390.

120. How many rods of fence will inclose 10 acres in the form of a square? *Ans.* 160 rods.

121. A farmer wishes to put out 100 bu. of lime per acre on a field, putting 1 bu. on a heap in regular rows; how many feet apart must the rows be? *Ans.* 20.87 ft.

122. C and D engage in trade with different sums of money; C loses 40% of his capital, and D gains 50% of his, when their capitals are just equal; how much greater was C's than D's when they began? *Ans.* $2\frac{1}{2}$ times.

123. A room is 20 ft. long, 16 ft. wide, and 12 ft. high; what is the distance in a straight line from one of the lower corners to the upper opposite corner? *Ans.* 28.28+ feet.

124. How many gallons of water will fill a circular cistern 6 ft. deep and 4 ft. in diameter? *Ans.* 564.019 gallons.

125. How deep must a circular cistern be made to hold 1000 gallons, if the diameter is 6 ft.? *Ans.* 4.728 feet.

126. I have sold 50 bu. of wheat for A, and 60 bu. for B, receiving for both lots $150; if A's wheat is worth 20% more than B's, how much ought I payeach?
Ans. $75.

127. Sold a cow for $30, by which I lost 16⅔%; afterwards sold a second at a gain of 10%, which exactly covered my loss on the first; what was the second cow sold for? *Ans.* $66.

128. If I lose 12% by selling sugar at $8 per cwt., what per cent do I gain or lose by selling it at $9?
Ans. Lose 1%.

129. The increase of a flock of sheep was 228, and this was 12% on the original number; what was the original number? *Ans.* 1900.

130. ⅔ of the price received for an article is equal to ¾ of its cost; what is the gain per cent? *Ans.* 12½%.

131. How much grain must I take to mill to bring away 2 bushels after the miller has taken 10 per cent for toll? *Ans.* 2¾ bu.

132. Sold tea at 30 cts. above cost and gained 16⅔%; what was the tea sold for? *Ans.* $2.10.

133. When gold was worth 50 per cent more than currency, what was the value in gold of a dollar bill?
Ans. 66⅔ cts.

134. A and B traded in company and gained $750, of which B's share was $600; A's stock was $1200; what was B's stock? *Ans.* $4800.

135. If 8 men cut 84 cords of wood in 12 days, working 7 hours a day, how many men will cut 150 cords in 10 days, working 5 hours a day? *Ans.* 24 men.

TEST EXAMPLES.

136. What is the difference between the true discount and the bank discount of $359.50 for 90 days, without grace, at 12%? *Ans.* $.315.

137. A man bought stock at 25% below par and sold it at 20% above par; what per cent did he make?
Ans. 60%.

138. What must I ask for cloth costing $4 a yard, that I may deduct 20% from my asking price and still make 20%? *Ans.* $6 per yd.

139. An agent received $502.50 to buy cloth, after deducting ½% commission; how many yards did he buy at $1.25 a yard? *Ans.* 400 yds.

140. A, B and C gain by trade $5200. A's stock was $2400, which was ¾ of B's, and B's was ⅔ of C's; what was the gain of each? *Ans.* A, $1200; B, $1600; C, $2400.

141. A and B take a contract of street-paving for $3300. A employs 50 men 20 days, and B, 40 men 30 days; how much shall each receive?
Ans. A, $1500; B, $1800.

142. A man engages 30 men, 40 women and 50 boys to do a piece of work; each man is to receive one-third more per day than a woman, and each woman one-fourth more than a boy. They receive $600 for the work; what does each man, woman and boy receive?
Ans. Each man $6⅔, each woman $5, each boy $4.

143. What is the present value of 3 notes of $1000 each, due in one, two and three years respectively, money worth 6%? *Ans.* $2683.71.

144. A, B and C engaged in business, gain $4800; A's stock was $4000, B's $6000, and C's gain, $1200; required C's stock and A and B's gain?
Ans. C's stock, $3333⅓; A's gain, $1440; B's, $2160.

145. A's stock was in trade 12 months, B's 15 months, and C's 18 months. A's gain was \$600, B's 600, and C's \$900; the whole stock was \$7000; what was each man's share? *Ans.* A, \$2500; B, \$2000; C, \$2500.

146. A merchant commenced business with \$20000; for three successive years he gained 20%, which was added to the capital of the previous year; what had he then? *Ans.* \$34560.

147. The standard of silver in the United States is $\frac{9}{10}$ pure, and that of England $2\frac{7}{9}$% purer; what is the standard of England? *Ans.* $\frac{37}{40}$.

148. 2 is what per cent of 4? *Ans.* 50%.

149. 12 is what per cent of 6? *Ans.* 200%.

150. 15 is what per cent of 50? *Ans.* 30%.

151. If A's money is 60% more than B's, how many per cent less than A's is B's? *Ans.* $37\frac{1}{2}$%.

152. Standard gold and silver in the United States is 9 parts pure and 1 part alloy; what per cent is alloy? *Ans.* 10%.

153. 25% of an article is how many per cent of $\frac{5}{6}$ of it? *Ans.* 30%.

154. If I sell $\frac{4}{5}$ of an article for what the whole cost, what per cent do I gain? *Ans.* 25%.

155. If I sell $\frac{3}{4}$ of an article for what $\frac{2}{3}$ cost me, what per cent do I lose? *Ans.* $11\frac{1}{9}$%.

156. The premium on a draft at $2\frac{1}{2}$% was \$25; what was the face of the draft? *Ans.* \$1000.

157. At what rate per cent simple interest will a sum of money double itself in 5, 6, 8 and 9 years respectively? *Ans.* 20%, $16\frac{2}{3}$%, $12\frac{1}{2}$% and $11\frac{1}{9}$%.

158. My annual interest at 6% is \$3000; what is my principal? *Ans.* \$50000.

159. In what time will a sum of money double itself, simple interest at 5%? 6%? 8%? and 10% respectively?
Ans. 20, 16⅔, 12½ and 10 years.

160. What is the rate of interest when $120 is paid for the use of $600 for 3 years and 4 months simple interest?
Ans. 6%.

161. What principal at 6% simple interest will amount to $6900 in 2 years and 6 months? *Ans.* $6000.

162. In what time will a sum of money treble itself at 5%? 6%? 8%? 10%, simple interest?
Ans. 40, 33⅓, 25, 20 years.

163. A steamboat whose rate is 12 miles an hour sails up a river whose current is 4 miles an hour, returning in 15 hours; how far up the river did the boat go?
Ans. 80 miles.

164. A and B engaged a piece of work. A can do it in 8 days, and B in 12 days; how long will it take them both to do it? *Ans.* 4⅘ days.

165. A can do a piece of work in ½ day, B in ⅓ and C in ¼ day; how long will it take them all working together to do it? *Ans.* ½ day.

166. A tank has two pipes for supply and one for discharge; the first alone will fill it in 9 hours and the second in 12 hours, and the third will discharge it in 8 hours; in what time will the tank be filled with all the pipes in operation? *Ans.* 14⅖ hours.

167. A boy driving geese to market, sold to his first customer one-half of what he had and one-half a goose more: to the second and third customer he sold in like manner, without killing a goose, when he had none left; how many had he at first?
Ans. 7 geese.

168. What is the cost of excavating a ditch 1 mile long, 6 feet wide, and 3 feet deep, at 40 cents per cubic yard? *Ans.* $1408.

169. A rectangular lot of land is 80 rods in length 40 rods in width; how many acres does it contain? *Ans.* 20 acres.

170. A triangular lot has one side 40 rods long, and a perpendicular line from the opposite corner of the lot to this side 16 rods; how much land does it contain? *Ans.* 2 acres.

171. A house valued at $15000, and furniture $3000, is insured for $\frac{3}{4}$ of its value at $1\frac{1}{4}\%$; what is the premium? *Ans.* $180.

172. What per cent on the value of a house, is a premium of $24 on $960, which is $\frac{1}{4}$ the value of the house? *Ans.* $1\frac{1}{4}\%$.

173. The principal is $900, the time 2 yr. 4 mo. 20 da., and the interest $150.50; what is the per cent? *Ans.* 7%.

174. The principal is $1350, the interest $225.75, the rate 7%; what is the time? *Ans.* 2 yr. 4 mo. 20 da.

175. Paid $850 in currency for gold at $6\frac{1}{4}\%$ premium; how much gold was received? *Ans.* $800.

176. Goods sold at 10% gain for $3300; what would have been the gain per cent if sold for $3600? *Ans.* 20%.

177. At what rate must I buy 7% bonds, in order to realize 10% on the investment? *Ans.* 70%.

178. In a sale of damaged goods for $1200, the loss was $300; what was the rate per cent? *Ans.* 20%.

TEST EXAMPLES.

179. Bought a farm for $12000, on condition that 15% is to be paid in cash, 23% in 6 months, 17% in 12 months, 30% in 18 months, and the balance in two years; how much was the last payment? *Ans.* $1800.

180. Sold goods for $2500, which was $300 more than cost; what was the gain per cent? *Ans.* $13\frac{1}{11}\%$.

181. What per cent of 35 is 3? *Ans.* $8\frac{4}{7}\%$.

182. What per cent of 3 is 35? *Ans.* $1166\frac{2}{3}\%$.

183. What is the amount of a 6% note for $1040, with interest from April 1st, 1875, to August 21st, 1878?
Ans. 1251.46\frac{2}{3}$.

184.   BALTIMORE, MD., Jan. 1st, 1876.
Three months after date, I promise to pay to John Smith or order, Five hundred dollars, for value received.
                                    JOHN JONES.

What was due on this note April 16th, 1877?
Ans. $531.25.

185.   BALTIMORE, MD., Jan. 1st, 1876.
Three months after date, I promise to pay Peter Post or order, One thousand dollars, for value received, with interest; how much interest is due on this note July 21st, 1877? *Ans.* $93\frac{1}{4}$.

186. What is the value of a note for $1200, due 2 yr. 6 mo. previous to this date, at 6%, simple interest? Compound interest payable annually and also semi-annually?
Ans. At simple interest, $1380; compound interest, annually, $1388.77; compound interest, semi-annually, $1391.13.

187. Date of note April 1st, 1872, for $1000, at 6%, payments July 16th, 1873, $200; April 1st, 1875, $300; April 16th, 1876, $400. What is due July 1st, 1877?
Ans. $331.57.

188. A note for $1000, dated April 1st, 1872, due in 3 years, with interest at 6%, payable annually, is endorsed as follows: Oct. 1st, 1873, $300; Jan. 1st, 1874, $200; what is due April 1st, 1875? *Ans.* $648.296.

189. A note for $500, dated April 1st, 1875, due in 3 years, with interest at 6%, payable annually, is endorsed: April 1st, 1876, $130; Oct. 1st, 1876, $100; Oct. 16th, 1877, $100. Jan. 1st, 1878, the note is discounted in bank, without grace, at 8%; what sum is received from the bank? *Ans.* $232.76.

NOTE.—In the last two examples, as the interest is only due at the end of the year, the payments made during the year should bear interest until the settlement at the end of the year, and this method is called the Merchants' Rule, which is decidedly the most just method.

190. What is the difference of the amounts of $2000 at 6% for 12 years at simple, and at compound interest?
Ans. $584.394.

191. What is the amount of $2000 at 6% simple interest for 20 years, also at compound interest?
Ans. Amount at simple interest, $4400; at compound interest, $6414.272.

192. What is the compound amount of $1000 for 20 years 3 months? *Ans.* $3255.243.

NOTE.—When there are months or days, in compound interest, take from the table for the years and on this compound for the months or days, as you would from the table; for 6 months multiply 1.03, when at 6%, and accordingly other times and rates.

193. The length of a rectangle is 128 feet and its breadth 96 feet; what is the length of its diagonal?
Ans. 160 feet.

194. The length of a rectangle is 72 feet and its diagonal 90 feet; what is its width? *Ans.* 54 feet.

195. A merchant of St. Louis has a debt of $6000 to pay in New York. Direct exchange is $1\frac{1}{2}\%$ premium on New York, but on New Orleans funds there is a discount of $\frac{1}{2}\%$, and at New Orleans the premium on New York is $\frac{3}{4}\%$; what advantage is the circuitous route?
Ans. 75.22\frac{1}{2}$ gained by the circuit.

196. A house valued at $12500, is insured for $\frac{4}{5}$ its value at 5%; for what sum is the policy drawn so as to include the premium? *Ans.* 10526\frac{4}{19}$.

197. If I purchase a city bond at 90% which bears 8% interest, what per cent do I get on my investment?
Ans. $8\frac{8}{9}\%$.

198. A merchant being asked at what per cent profit he sold his goods, replied $\frac{4}{5}$ of my selling price is 25% less than cost; at what per cent profit did he sell goods?
Ans. 12% profit.

199. In the ruins of Persepolis are two columns standing upright; the one is 70 feet above the plane and the other 50 feet; in a straight line between these columns is a small statue 5 feet high, the head of which is 100 feet from the top of the higher and 80 feet from the top of the lower column. What is the distance between the tops of the columns? *Ans.* 143.54.

200. What is the rate per cent, when the amount is $\frac{5}{4}$ the principal that had been at interest 3 years?
Ans. $8\frac{1}{3}\%$.

TEST EXAMPLES.

201. How much more fence will it take to enclose a rectangular lot of land 100 rods long and 36 rods wide, than if the same quantity of land were in the form of a square ? *Ans.* 32 rods.

202. In what direction and at what rate must a man at the equator, having the sun on his meridian, travel, in order that he may always have noon?
Ans. He must travel west $1037\frac{1}{4}$ miles per hour.

203. Extract the square root of .0043046721.
Ans. .06561.

204. When it is 5 A.M. at the Cape of Good Hope, in longitude 18° 24' east, what is the time at Cape Horn, in longitude 67° 21' west?
Ans. 11 hr. 17 min., P.M., of the previous day.

205. How do you reduce several fractions to equivalent fractions having the least common denominator.
Ans. By multiplying both terms of each fraction by the quotient obtained by dividing its denominator into the least common multiple of all the denominators.

206. How many steps of three feet each would a man take in walking five miles ? *Ans.* 8800 steps.

207. The difference in the time of two places is 2 hr. 10 min.; what is the difference of longitude?
Ans. 32° 30'.

208. If 500 copies of a book of 210 pages take 12 reams of paper, how much will 1000 copies of 280 pages require ? *Ans.* 32 reams.

209. Find the third term of 3 : 8 :: ? : 12.
Ans. $4\frac{1}{2}$.

210. If $2\frac{1}{4}$ yd. cloth cost $20, what will 18 yd. cost ?
Ans. $160.

211. Find the greatest common divisor of 492, 744, 1044. *Ans.* 12.

212. Find the least common multiple of 25, 36, 33, 12, 45. *Ans.* 9900.

213. When the extremes and the number of terms are given in a series of equal differences, how do you find the sum of the series?

Ans. Multiply the sum of the extremes by one-half the number of terms.

214. Of what factors of several numbers is the greatest common divisor composed?

Ans. The factors common to all the numbers.

215. When are four quantities in proportion?

Ans. When the quotient of the first and second, and that of the third and fourth are equal.

216. How is the rate per cent obtained, when the principal, interest, and time are given?

Ans. Divide the interest by the product of the principal and time.

217. How do you find the time, having given the principal, interest, and rate?

Ans. Divide the interest by the product of the principal and rate.

218. Having given the principal, rate, and time, how do you find the interest?

Ans. The product of the principal, rate, and time is equal to the interest.

219. How do you find the principal, when the interest, rate, and time are given?

Ans. Divide the interest by the product of the rate and time.

220. How do you find the last term of a series of equal ratios, the first term, ratio, and number of terms being given?

Ans. The last term is equal to the ratio raised to the power expressed by the number of terms less one, multiplied by the first term.

221. How do you find the sum of a series of equal ratios?

Ans. The sum is equal to the product of the ratio and last term, diminished by the first term, and divided by the ratio less one.

222. How much will it cost to dig a cellar 60 feet long, 30 feet wide, and 6 feet deep, at 30 cts. per cubic yard? *Ans.* $120.

223. A and B enter into business; A furnishes $600 for 8 months and B $400 for 9 months; they gain $140; what is each one's share? *Ans.* A's $80, B's $60.

224. 5 pence is what per cent of $8\frac{1}{3}$ shillings?
Ans. 5%.

225. What is the greatest common divisor of $\frac{1}{2}$, $\frac{2}{3}$, $\frac{3}{4}$, $\frac{4}{5}$? *Ans.* $\frac{1}{60}$.

226. What is the least common multiple of $\frac{1}{2}$, $\frac{2}{3}$, $\frac{3}{4}$, $\frac{4}{5}$? *Ans.* 12.

227. What is the least common multiple of 3, 4, 5, 6, 7, 8? *Ans.* 840.

228. How is a common fraction reduced to a decimal?
Ans. Perform the division indicated.

229. Find the term omitted in 4 : ? :: 9 : 16?
Ans. $7\frac{1}{3}$.

230. What principal will gain $11.20 in 3 years and 6 months at 8%? *Ans.* $40.

231. How do you change a decimal into a common fraction?

Ans. Make the decimal the numerator, and for the denominator place 1 with as many zeros annexed as there are figures in the decimal.

232. What is the least common multiple of 21, 35, 42? *Ans.* 210.

233. How many yards of cloth, ⅝ yd. wide, are equivalent to 24 yards, ⅞ yd. wide? *Ans.* 22⅕ yd.

234. The difference between ⅓ and ⅝ of a number is 8; what is the number? *Ans.* 576.

235. The first term of a series of equal differences is 1, the difference 2, and the number of terms 8; what is the sum of the terms? *Ans.* 64.

236. The first term of a series of equal ratios is 1, ratio 3, and the number of terms 6; what is the sum of the series? *Ans.* 364.

237. A, B, and C bought a horse for $120 and sold him for $180. A received for his part of the gain $25, and B $20; how much had each paid for the horse?
Ans. A $50, B $40, C $30.

238. Find the sum of 8 terms of the series of equal ratios 3, 6, 12, etc. *Ans.* 765.

239. A body of 3600 troops have ⅕ as many cavalry as infantry; what is the number of each?
Ans. 600 cavalry and 3000 infantry.

240. When snow is uniformly 1 foot deep, how many cubic yards on an acre of land?
Ans. 1613⅓ cubic yards.

241. To what number must 235 be added three times to make 950? *Ans.* 245.

TEST EXAMPLES.

242. The product of three factors is $19\frac{1}{4}$; two of the factors are $5\frac{1}{4}$ and $1\frac{1}{4}$; what is the third factor?
$Ans.$ $2\frac{13}{14}$.

243. How much gold will $500 currency buy, when gold is worth $1.25? $Ans.$ $400.

244. If 12 men in 5 days of 8 hours plough 60 acres of land, how many men will plough 100 acres in 8 days of 10 hours each? $Ans.$ 10 men.

245. What is the bank discount of $600 for 3 months at 7%? $Ans.$ $10.85.

246. What is the amount of $600 at 6% compound interest, for 10 years, payable annually? Payable semi-annually?
$Ans.$ Annually, $1074.51; semi-annually, $1083.67.

247. What principal at 6% will amount to $960 in 2 yr. 6 mo., simple interest?
$Ans.$ $960 \times \frac{100}{115} = \$834.78\frac{2}{23}$.

248. A man owning $\frac{2}{3}$ of a bank, sold 30% of his share; what per cent of the whole had he left?
$Ans.$ $46\frac{2}{3}\%$.

249. Add 100.01, 1000.001, 10000.0001, 12.004, 10.0004, 9.00003. $Ans.$ 11131.01553.

250. The means of a proportion and one extreme; thus, $3:4::9:?$. $Ans.$ 12.

251. The extremes and one mean given; thus, $2:6::?:9.$ $Ans.$ 3.

252. One couplet of a ratio and one term of another, to find the second term; thus, $\frac{2}{3}$ and $\frac{1}{10}$? $Ans.$ 15.

COR. 1.—When the two means and one extreme are given, divide the product of the means by the given extreme.

TEST EXAMPLES. 273

COR. 2.—When the two extremes and one mean are given, divide the product of the extremes by the given mean.

COR. 3.—When one couplet, forming a ratio, is given, and one term of another couplet, to form an equal ratio: The ratio of the unknown to the similar term must be the same as of the given similar terms.

253. A house valued at $6000 is insured for $\frac{4}{5}$ its value at $\frac{5}{8}\%$; what is the premium? *Ans.* $30.

254. A clerk at a salary of $1000, pays $250 for board, $175 for clothing, and $225 for other expenses; what per cent of his salary is left? *Ans.* 35%.

255. Two merchants engage in business; the one puts in $8000 and the other $6000; the latter is to receive $500 extra annually for his attention to the business; if the gain is $2800 annually, what is each one's share?
Ans. First, 1314\frac{2}{7}$; and second, 1485\frac{5}{7}$.

256. Bought 200 lb. bacon at 8 ct. per lb., and when I sold it at 10 ct. per lb. I found it had lost in weight $\frac{1}{4}$; did I gain or lose and how much? *Ans.* Gained $1.50.

257. What will it cost to enclose in one square field 160 acres of land, the fence costing $1.50 per rod?
Ans. $960.

258. The premium for the insurance of property for $12000 was $90; what was the rate per cent?
Ans. $\frac{3}{4}\%$.

259. A man has 50% of his property invested in a farm, 10% in a house, 15% in stock and farming implements, and the balance $3000 in bank; what is the amount of his property? *Ans.* $12000.

260. If 40 acres of land form a square, what is the length of a side? *Ans.* 80 rods.

261. If 80 acres are in the form of a rectangle, the length being double the width, what are the dimensions? *Ans.* Length 160 rods, breadth 80 rods.

262. Bought a carriage for $258.75, which was sold at a profit of 15%; what was the original cost?
Ans. $225.

263. The difference in the time of St. Petersburg and Washington is 7 hr. 9 min. 19 sec.; what is the difference in longitude? *Ans.* 107° 19′ 45″.

264. In what terms of a proportion may equal factors be cancelled without altering the ratios?

Ans. In both terms of the same couplet, in both antecedents, or in both consequents.

265. How is a proportion formed with only three numbers?

Ans. By repeating the middle number; thus, as 2 : 4 :: 4 : 8; and the 4 is said to be a mean proportional to the other two terms.

266. What is the value of a fraction multiplied by its denominator?

Ans. The numerator becomes an integral number.

REM.—In Fahrenheit's thermometer, the freezing point of water is 32° and the boiling point 212°; in the Centigrade, the freezing point is 0° and the boiling 100°; in Raumer's, the freezing is 0° and the boiling 80°.

267. If Fahrenheit is at 86°, what will C. indicate?
Ans. 30°.

268. If Fahrenheit is at 77°, what will R. indicate?
Ans. 20°.

NOTE.—212 − 32 = 180; the ratios will be formed by the numbers 180, 100, 80, and will be $C = \frac{5}{9} F$, $R = \frac{4}{9} F$, $C = \frac{5}{4} R$, $R = \frac{4}{5} C$, $F = \frac{9}{5} C$, and $F = \frac{9}{4} R$.

REM.—Two bodies attract each other inversely as to their weights.

269. Suppose the earth and moon were not interrupted by any other attractions, their weights being as 49147 and 123 respectively; how many miles would the moon move whilst the earth moved 100 miles?

Ans. 39956$\frac{112}{123}$ miles.

270. At what time between 3 and 4 o'clock are the hour and minute hands together?

Ans. 3 hr. 16$\frac{4}{11}$ min.

REM.—Whilst the hour hand goes the space of 5 minutes, the minute hand goes the round of the dial 60 minute spaces; the gain is $\frac{55}{60} = \frac{11}{12}$; that is, in 1 hour it gains 55 minutes, or it gains 11 in 12.

At 3 o'clock the hands are 15 minute spaces apart, and this is to be gained until they are together; therefore, as 11 gain : 3 gain :: 12 hr. : 3 hr. 16$\frac{4}{11}$ min.

271. At what time between 4 and 5 is the minute hand 5 minute spaces behind the hour hand?

Ans. 4 hr. 16$\frac{4}{11}$ min.

REM.—The gain is the same as in 270.

272. At what time between 4 and 5 is the minute hand 5 minute spaces ahead of the hour hand?

Ans. 4 hr. 27$\frac{3}{11}$ min.

273. The diameter of the earth is 7924 miles; what is its circumference? *Ans.* 24894.0384.

274. What is the length of a degree of a great circle?

Ans. 69.15 + miles.

275. The diameter of the Arctic and Antarctic Circles is 2625.5 miles; what is the length of a degree of longitude there? *Ans.* 22.9 + miles.

REM.—At the poles all the degrees of longitude come to a point.

276. A's money is 10% of B's, and $12\tfrac{1}{2}$% of C's; B has $50 more than C; how much has each?
Ans. A has $25, B $250, and C $200.

277. If a globe of gold $\tfrac{1}{4}$ in. in diameter is worth $15, what is the value of a globe 4 in. in diameter?
Ans. $61440.

278. How far distant is a cloud in which lightning is seen 15 seconds before the thunder is heard?
Ans. 16500 feet.

NOTE.—Sound moves 1100 feet and light 192500 miles in a second.

279. What is the value of one dollar U. S. in English money? *Ans.* 4s. 1$\tfrac{1}{4}$d.+.

NOTE.—24 carats gold is pure gold; 18 carats has 6 carats alloy, 16 carats has 8 carats alloy, etc.; the alloy is generally copper and silver.

280. A refiner melts 8 lb. gold 20 carats fine with 12 lb. 22 carats fine; how much alloy must be added in order to make it 18 carats fine? *Ans.* 3$\tfrac{1}{3}$ lb.

281. 140 is 20% less than what? *Ans.* 175.

282. 168 is 20% more than what? *Ans.* 140.

283. 540 is 33$\tfrac{1}{3}$% more than what? *Ans.* 405.

284. A lost $5, which was 25% of what he had left; how much had he at first? *Ans.* $25.

285. Sold 324 bushels of wheat, which was 33$\tfrac{1}{3}$% of what I had left; how many bushels in the whole crop?
Ans. 1296 bushels.

286. If 20% of 37½% of 33⅓% of a number is 4, what is the number? *Ans.* 160.

287. If the cost of a coat is $30, the trimmings cost 60% less, and the making 40% less than the cloth, what did each cost ? *Ans.* Cloth $15, trimming $6, making $9.

288. If I sell ⅔ of an article for what the whole cost, what per cent do I gain? *Ans.* 50%.

289. If I sell ⅚ of an article for what ¾ cost, what per cent do I gain ? *Ans.* 11⅑%.

290. What is the cost of grading 2 miles of railroad, averaging 3 feet in depth or height, and 12 feet wide, at 30 cents per cubic yard ? *Ans.* $4224.

291. I sold goods through a broker who charged me 2¼%; my commission was 4¾%; the transaction netted me $50; for what were the goods sold ? *Ans.* $2000.

292. A hare is 75 leaps ahead of a hound, and the hare takes 2 leaps while the hound takes one, but two of the hound's leaps is equal to five of the hare; how many leaps must the hound take to overtake the hare?
Ans. 150 leaps.

NOTE.—Whilst the hound takes two leaps the hare takes four leaps, but the two leaps of the hound is equal to five of the hare; hence the hound gains 1 hare leap on the hare for every two leaps the hound takes.

293. A tank is 24 ft. long, 8 ft. wide, and 9 ft. high; how many bushels of wheat will it hold ?
Ans. 1388½ + bu.

294. How much currency at 12½% discount, will purchase 50 shares ($100 each) of bank stock at a premium of 12% ? *Ans.* $6400.